Synthesis Lectures on Engineering, Science, and Technology

The focus of this series is general topics, and applications about, and for, engineers and scientists on a wide array of applications, methods and advances. Most titles cover subjects such as professional development, education, and study skills, as well as basic introductory undergraduate material and other topics appropriate for a broader and less technical audience.

Tadahiro Kuroda

The Super-Evolution
of Semiconductors

 Springer

Tadahiro Kuroda
University of Tokyo
Tokyo, Japan

ISSN 2690-0300   ISSN 2690-0327   (electronic)
Synthesis Lectures on Engineering, Science, and Technology
ISBN 978-3-031-60517-8   ISBN 978-3-031-60518-5   (eBook)
https://doi.org/10.1007/978-3-031-60518-5

Translation from the Japanese language edition: "The Super-Evolution of Semiconductors" by Tadahiro Kuroda, © Tadahiro Kuroda 2023. Published by Nikkei Business Publications, Inc. All Rights Reserved.

This Springer imprint is published by the registered company Springer Nature Switzerland AG
The registered company address is: Gewerbestrasse 11, 6330 Cham, Switzerland

If disposing of this product, please recycle the paper.

To Mari, Eri, and Yoshito

Preface

All of a sudden, the frog jumped out of the boiling water!

I wonder if that is how Japan's latest semiconductor policy looks to the world.

In the last quarter century, Japan's semiconductor industry was indeed like a boiling frog. Since reaching 50% in 1988, the share of the world's semiconductor market held by Japanese companies collectively has been falling, dropping to a mere 10% today. Despite several policy attempts to stem the loss, the decline persisted. Japan's policy continued to be criticized as being too little, too late.

However, suddenly, the frog jumped.

Rapidus' manufacturing process goal is beyond 2 nm. Japan is determined and putting serious efforts into reviving our semiconductor industry. This time around, the world is surprised by the scale and speed of our policy.

So, what is happening in Japan? That is what this book is about.

Another theme of the book is the future of the industry.

The industry has been engaging in increasingly fierce capital competition, driven by a greedy strategy where bigger is better. Can that continue into the future? Instead of fighting for supremacy, isn't it possible to promote democratization so as to create diverse products to drive the world toward prosperity?

The hints for the answer can be found in the diversity of nature. More than 160 years after Darwin proposed his theory to explain the evolution of nature, the latest science is attempting to unearth the hidden evolutionary mechanism of living organisms. Specifically, while competing fiercely for survival, life forms have been intricately connected across species, helping each other to evolve. The key to realizing such co-existence and co-evolution has been agility.

This book expounds on the super-evolution of semiconductors.

Since the release of the original Japanese version of the book in May 2023, there have been discussions of publishing translated versions in China, Taiwan, and Korea. Thanks to Professor Anantha Chandrakasan, Dean of the School of Engineering at Massachusetts

Institute of Technology, for introducing me to Mr. Charles B. Glaser of Springer Nature, I am now able to make the book available in English as well. I would also like to thank my colleagues Wai-Yeung Yip and Shogo Kondo for translating the Japanese manuscript into English.

Tokyo, Japan   Tadahiro Kuroda
November 2023

Contents

1.1 Banquet: The Stage Turns

On the night of December 14, 2022.

The Hotel Okura's main banquet hall, the Heian-no-Ma, was filled with the enthusiasm of 400 people. The guests were finally seated when the host's avatar appeared on a large screen.

VIPs lined up for front-row center seats.

Akira Amari, Chairman of the Liberal Democratic Party Parliamentarians for Semiconductor Strategy Promotion; Yasutoshi Nishimura, Minister of Economy, Trade and Industry (METI); Makoto Gonokami, President of RIKEN. Also from METI: Satoshi Nohara, Director General, Commerce and Information Policy Bureau, Kazumi Nishikawa, Director, General Affairs Division, Hisashi Kanazashi, Division Director, and Yohei Ogino, Director.

From industry: NTT Chairman Jun Sawada, JSR Honorary Chairman Mitsunobu Koshiba, Tokyo Electron former Chairman Tetsuo Tsuneishi and President Toshiki Kawai, Advantest President Yoshiaki Yoshida, SCREEN Holdings President Toshiro Hiroe, Sony Semiconductor Solutions President Terushi Shimizu, Kioxia President Nobuo Hayasaka, Hidetoshi Shibata, President of Renesas Electronics, Nobuaki Kawahara, Director of Mirise Technologies, Akihiro Teramachi, President of THK, and Atsushi Horiba, Chairman and CEO of Horiba Mfg.

Also in attendance were Director Dario Gil and Deputy Director Mukesh Khare of IBM Research Laboratories; Yaw Debock, Chief Strategy Officer, imec; Ajit Manocha, President, SEMI (Semiconductor Equipment and Materials International); Jun Ho, Vice President, TSMC; Makoto Onodera, President, TSMC Japan; and TSMC Japan 3DIC Research and Development Center Director Yutaka Emoto.

T. Kuroda, *The Super-Evolution of Semiconductors*, Synthesis Lectures on Engineering, Science, and Technology, https://doi.org/10.1007/978-3-031-60518-5_1

Then Chairman Tetsuro Higashi and President Atsuyoshi Koike of Rapidus, who were the stars of the evening, took their seats in the center.

Everyone was filled with optimism for the future.

The semiconductor industry is a growth industry.

The semiconductor market, which was $15 billion in 1982, reached $500 billion in 2021. The high growth rate of 9.4% per year on average has continued for 40 years.

Initially, the semiconductor market was about 0.2% of the nominal GDP. Then, in the mid-1990s, the market grew rapidly to 0.4%. What happened?

Many people may recall the events of the mid-1990s when Windows 95 became a worldwide hit. Prior to that, semiconductors were used in many consumer electronics products that enriched the physical space, such as televisions and video players, but since then they have been used in personal computers (PCs) and smartphones.

The PC created the virtual space, while the smartphone made it portable.

In other words, the semiconductor market grew from 0.2% to 0.4% of GDP as the application of semiconductors expanded from the physical to virtual space (cyberspace).

In recent years, the semiconductor market seems to be on the verge of another surge toward 0.6% of GDP.

We will have to watch the market a little more carefully before drawing a conclusion because of the special demand caused by the COVID-19 pandemic. The market has already entered an adjustment phase. However, if significant growth again emerges beyond the adjustment phase, semiconductors will enter a third period of growth.

The new value created by semiconductors is the creation of a data-driven society through the high-degree fusion of physical and virtual space, which will both solve social issues and promote economic development.

Automated driving, robotics, and smart cities are just a few examples. Real-time data in the physical space is collected by sensors, and then they are analyzed by AI (artificial intelligence) using a digital twin in the virtual space, with the results immediately fed back to the physical space to create action through motor control. Thus, people can travel safely and comfortably to their destinations in the shortest time and with the least energy.

The world's semiconductors are growing robustly in this way.

On the other hand, looking at Japan domestically, the Japanese semiconductor industry has been in a slump for the past quarter century. While Korea, Taiwan, and China have been growing rapidly, Japan has been in stagnation. Various factors have been identified as reasons for the decline of the Japanese semiconductor industry. These include business environment factors such as Japan-US trade frictions and a strong yen, strategic factors such as delays in digitalization and shift to a horizontal specialization business model [1], and policy factors such as Japan's self-reliance policies and its inability to counter policies of South Korea, Taiwan, and China to foster national enterprises.

But the tide has turned.

The business environment has improved with positive developments in Japan-US cooperation and a weaker yen. In addition, investment strategies in digital industries and foundries have turned aggressive. And industrial policy has also turned to promote international collaboration.

The nation is determined, that is, it is putting its national fortunes on the line and making a serious effort to revitalize the semiconductor industry.

This time it will be different.

The theme of SEMICON Japan 2022, an exhibition organized by SEMI, was "Change the future. The future is changing."

Prime Minister Fumio Kishida arrived at the opening ceremony and issued the following proclamation.

"Semiconductors, needless to say, are the key technology behind digitization, decarbonization, and economic security [2].

They are also the most important resource supporting the new capitalism, which aims to transform social issues, such as green and digital transformations, into engines of growth and to realize a sustainable economic society.

We will overcome the COVID-19 pandemic, promote the normalization of socioeconomic activities, and take advantage of the weak yen. To that end, we will support the aggressive expansion of domestic investment in semiconductors, which are needed by society, and increase the resilience of the economic structure.

It is estimated that TSMC's semiconductor plant in Kumamoto will have an economic impact of over 4 trillion yen (about $28.6 billion at a rate of $1 = 140 yen) and create over 7,000 jobs in the region over 10 years.

In order to encourage such investments, which will revitalize local regions, the recently enacted supplementary budget provides for 1.3 trillion yen (about $9.3 billion) [3]. This will allow us to expand domestic investment in semiconductors nationwide and promote the development of next-generation semiconductors.

We must also face the reality that it is difficult to establish the semiconductor supply chain in just one country. Therefore, we will also strengthen global collaboration in government-supported semiconductor development projects.

Yesterday, Rapidus, the mass production center for future next-generation semiconductors, announced the signing of a joint development partnership with IBM. In addition, the company aims to achieve mass production in the late 2020s in cooperation with imec in Europe [4].

We hope to supply cutting-edge semiconductors from Japan to support the digital economy, which will greatly evolve globally in the future, including advanced computing systems such as AI and quantum computing, automated driving, and next-generation robots."

1.2   The University of Tokyo Made Its Move: Be Agile!

President Gonokami took the stage in the Heian-no-Ma banquet hall. Gonokami brought with him the strategies for semiconductors and quantum computing when he transitioned his role from president of the University of Tokyo to president of RIKEN.

It was in May 2019 when I met President Gonokami at the University of Tokyo. The main entrance to Yasuda Auditorium on the Hongo campus, covered in the vibrant green color of spring, was on the third floor above ground level. After showing my ID to the guard and entering the building, I went down the stairs and counterclockwise down the hallway to find myself in what looked like a secret military base.

President Gonokami asked me what was needed for the revival of the semiconductor industry in Japan.

"It is technology to efficiently develop and integrate energy-efficient specialized chips in 3D," I replied.

And I elaborated further.

"What is required to create a data-driven society, Society 5.0, is advanced computing [5]. This is Japan's most essential resource, along with energy.

The challenge in computing is how to improve energy efficiency. At the current rate, data centers' power consumption will skyrocket tenfold in 10 years. Without a solution to this energy crisis, there will be no sustainable development of a data-driven society.

The energy crisis is actually caused by AI. In order to perform ever-more sophisticated analysis of exploding data, the computational power of AI has increased by four orders of magnitude over the past decade. On the other hand, the power efficiency of the general-purpose processors used for that computation has improved by only one order of magnitude over the same decade [6].

The technology to improve energy efficiency lies in semiconductor scaling and 3D integration. In terms of scaling, Japan has fallen far behind the world's leading edge, and so we need to learn from overseas to catch up.

On the other hand, Japan has a number of excellent technologies for the materials and manufacturing equipment needed for 3D integration, and we should utilize this strategic advantage. If 3D integration can reduce the distance of data movement to a different order of magnitude, energy consumption in moving data will be greatly reduced [7].

Since the packaging process for 3D integration requires substantially smaller investments than wafer process scaling, 3D integration has a higher return on investment.

On the other hand, with regard to design technology, specialized chips that have been stripped of unnecessary circuitry can reduce energy consumption by orders of magnitude compared to general-purpose chips. The movement to develop specialized chips has already begun at companies such as GAFA and Tesla. The era of general-purpose chips has been one of capital competition, but the era of specialized chips will be one of knowledge competition. In other words, innovation in design and development is required.

The development of specialized chips is becoming increasingly difficult. In recent years, it takes one year and costs $100 million even with 100 designers. With such long development time and large expenses, interest in specialized chips has been waning and the design population shrinking. Indeed, Japan's industry has been losing interest and capacity in specialized chips development. Under such an environment, even if it is able to build new plants and expand its manufacturing capacity, it will not be able to use it to strengthen its industry immediately.

AI technology is evolving rapidly, and the processing software is updated on a monthly basis. In the digital economy, the key is to highly integrate hardware and software to create innovation and iterate improvements in rapid cycles. But this becomes difficult when the development speeds of hardware and software are so far apart. However, a silicon compiler that can compile a program to automatically design a chip would make it possible to develop hardware quickly (and agilely).

Of course, the performance of an automatically designed circuit may achieve only 80% of the performance of a circuit that a designer took time to optimize, but that can be acceptable. This is the so-called 80-point principle. We obtain added value in increasing development efficiency by a factor of five in taking advantage of the 80–20 rule which states that 20% of the effort generates 80% of the outcome.

In addition, design assets must be reused to avoid explosive growth in the scale of the design project. Chiplets will become more important. 3D integration will be a strategic advantage in this regard as well, since the chiplets can be combined to form a complete system in a package."

The University of Tokyo sprang into action.

First, the university established d.lab in October 2019 to pursue open collaboration on campus. The "d" in d.lab embodies its mission to research the design of domain-specific systems, starting with data and extending from software to devices, for the age of digital inclusion when digital technology allows each individual to shine. In addition, to facilitate technology transfer to the younger generation, the office was built in a dormitory.

Next, in November, the university announced the world's first organization-wide joint research alliance on semiconductor technology with TSMC. At the joint press conference, Mark Liu, Chairman of TSMC, Philip Wong, Research Director and Professor at Stanford University, Dr. Makoto Gonokami, then President of the University of Tokyo, and Teruo Fujii, then Vice President and current President, joined hands in a show of commitment to the alliance.

The following year, in August 2020, the university established RaaS, a technical research association for closed industry-government-academia collaboration under strict confidential information management. In addition to being an abbreviation for Research Association for Advanced Systems, the name reflects its goal to provide "Research as a Service."

Currently, d.lab has 49 corporate sponsors, while RaaS has so far attracted a total of 12 participating companies. The goal for d.lab and RaaS is $10\times$ energy efficiency and $10\times$ development efficiency.

That is the same goal as Rapidus.

But the approach is complementary.

Specifically, Rapidus pursues scaling while the University of Tokyo turns to 3D integration to improve energy efficiency. In addition, Rapidus will shorten the manufacturing cycle while the University of Tokyo will reduce the design cycle to improve development efficiency.

1.3   More People: Attracting the World's Brains

In addition to technology, human resource development is an urgent issue. **Technology is created by people.**

For that reason, we started Agile-X, a semiconductor democratization center, in April 2022.

Agile means "quick" and "nimble."

Build a development platform that reduces the development time and cost of specialized chips to one-tenth and the world's brightest minds will come. As a result, we can democratize semiconductors by increasing the population of specialized chips designers tenfold. That is the goal of Agile-X.

Behind the goal of democratization is the idea of the "collective brain," that innovation is accelerated by the collision of ideas from many people.

For example, on the South Pacific islands, there is a strong correlation between the variety of fishing gears used and the island population. In other words, the more populated an island is, the more variety of gears is used.

Similarly, despite having a smaller brain than Neanderthals, Homo sapiens are believed to have invented and utilized a larger variety of tools due to their ability to form larger social groups, resulting in their evolutionary success.

If more people can develop their own chips, more innovation should occur. Based on that idea, a movement to democratize semiconductors has quietly begun around the world.

TSMC Chairman Mark Liu concluded his keynote address to the 2021 International Solid-State Circuits Conference (ISSCC) with the following words.

It is extremely important to foster a broad design ecosystem that lowers the barrier to entry and unleashes an enormous amount of hardware innovation. Ideally, it should be as easy to innovate in hardware as it is to write software codes.

Innovation comes from the free flow of ideas. Ideas come from people. Therefore, innovation becomes democratized when more people can create their own semiconductors. And as Dr. Liu noted, "When that happens, we will see another renaissance in application and system design."

In 1959, physicist Richard Feynman gave a lecture where he famously said, "There's plenty of room at the bottom," referring to the potential of nanoscale research. That sparked a global pursuit of micro devices. Eventually, microelectronics was born, which evolved into nanoelectronics.

Even as the traditional approach of scaling is reaching its limits, ongoing research known as More Moore continues to pursue further scaling to extend Moore's Law.

More recently, research has begun on "More than Moore," which aims to create new value-added as an alternative to scaling. 3D integration research is attracting global investment partly because of its high return on investment.

Borrowing Feynman's words, I believe "There's plenty of room at the TOP."

We are now approaching an integration level of more than 100 billion transistors on a chip. Intel CEO Pat Gelsinger predicted that by 2030, we can integrate a trillion transistors in a single package.

To enable more people to join forces to foster innovation, research on "More People" is crucial.

The reason that only giant corporations can afford to develop specialized chips is the result of the manufacturing ecosystem being optimized for mass production to serve an industrialized society.

In a knowledge-based society, time performance will be more important than cost performance. Since "time is money," time performance also encompasses cost performance.

Researchers know the value of being agile better than anyone else, since research publication is a race against time. It is important for researchers to have easy access to semiconductors tailored to their needs to help them contribute to the advancement of science.

It is therefore essential to promote the democratization of semiconductors in order to attract the world's brains. In other words, the goal is not to compete for the pie but to grow it.

Human resource development is the role of academia. Human resources are the capital of Japan that will create its future.

Japan can learn More Moore from the world and contribute to the world with More than Moore and More People.

And Japan will need to construct an advanced computing infrastructure. By combining conventional bits with qubits and neurons and integrating software with hardware, Japan will develop massive computing power and create a communication network to make it accessible from all over the country. This infrastructure for a digital society is what Japan should mobilize the country to realize.

1.4   The Semiconductor Forest: Co-existence and Co-evolution

Semiconductors are a strategic resource. Who will win the far-reaching competition for technological dominance? Geopolitical risks are rising, and the global outlook is becoming increasingly uncertain.

Instead of fighting for dominance, why not promote democratization and make semiconductors a common asset of the world, a global commons, so as to stimulate the creation of diverse chips to help bring prosperity to the world?

I feel that the clue to the answer lies in the diversity of life on Earth.

Before the Cretaceous Period (from approximately 145 to 66 million years ago), there were only one-tenth as many species of living organisms as today.

However, it all changed with the emergence of flowering plants [8].

Flowers relied on insects to transport pollen instead of directly spreading it. This caused a major shift in the relationship between plants and insects: instead of only serving as food for insects, plants could now benefit from insects too.

Flowers used vibrant colors to compete for the attention of insects. In turn, insects enhanced their ability to fly in response to changes in the shape of flowers. In this way, they promoted evolution in each other, resulting in repeated cycles of co-evolution.

Thus, forests became richer, mammals that fed on insects attracted to the flowers diversified, and primates evolved while consuming fruits produced from the flowers.

Eventually, flowers acquired a new ability.

It was the acceleration of their life cycle. The time from pollination to fertilization was reduced from one year to a few hours. This accelerated the evolution of all living organisms.

$$y = a(1 + r)^n$$

This is the formula for compound interest calculation, where (r) is the interest rate and (n) the number of times interest is compounded. Even if the initial investment (a) is small, the future value will increase significantly with a long-term investment.

Replacing (n) with (1/t) yields the basic formula for the digital economy, where (t) is the development cycle time. This equation applies to both chip performance improvement and company growth.

In other words, repeating improvements over and over again in rapid cycles is a growth strategy for the digital economy. Therefore, it is essential to increase the number of improvement cycles (n) rather than the improvement rate (r), i.e., to shorten the development cycle time (t).

That is why agile is the keyword.

While living organisms compete fiercely for survival, they are intricately interconnected across species and help each other to survive. In other words, they co-exist and co-evolve. And being agile is the key.

If we replace plants with chips, insects with chip users, and forests with ecosystems, then what do we have? What would be the "flower" of semiconductors that turns competition into co-existence and co-evolution?

The banquet was in full swing in Heian-no-Ma.

The avatar of the emcee invited violinist Taro Hakase to the stage. Hundreds of billions of semiconductor switches in the audience's smartphones were turned on and off hundreds of millions of times [9].

Taro Hakase played the Japanese song "Another Sky."

"There are also a lot of semiconductors used in airplanes," said Hakase, eliciting laughter from the audience.

But it is a different song—"Rhapsody in Blue"—that usually makes me think of flying. However, this time, it brought back memories of my hectic overseas business travel schedule in the past two months.

September 19: Discussion with President Luc Van den Hove at imec, Belgium.
September 24: Discussion with Dario Gil, Director, IBM Research Center, New York.
September 26: Discussion and ideas exchange with Mukesh Khare and others of IBM at the NanoTech Complex in Albany.
September 28: Research ideas exchange at Princeton University.
October 5: Research ideas exchange at UC Berkeley (University of California at Berkeley).
October 6: Research ideas exchange at Lawrence Berkeley National Laboratory.
October 10–14: Participation in a Japanese government delegation to the U.S. to visit the Department of Commerce (DOC).
October 31: Participation in a panel discussion on human resource development at SEMI's international conference ITPC.
November 7: Announcement of a collaboration between d.lab and imec at imec Technology Forum.

November 29: Visit to the TSMC Japan 3DIC R&D Center in Tsukuba with d.lab corporate sponsors.

Over the past two months, I met many times with the people gathered in Heian-no-Ma this evening to exchange ideas, in response to an urgent need to build a network for cooperation.

Hisashi Kanazashi, a METI section chief sitting next to me, said,

"I need to go to a meeting with the DOC now, so please excuse me."

He got up, and I said, "Thank you for your hard work. Good night."

Taro Hakase was playing "Jounetsu Tairiku" in front of the audience [10].

The future was finally about to start.

Regaining Lost Ground: Game Changer

2

2.1 Semiconductor Strategy: Learning from Kendo

Game changer

In June 2021, Japan's Ministry of Economy, Trade and Industry (METI) released its strategy for the country's semiconductor industry. One of the documents, entitled "Japan's Decline," showed that the global market share of Japanese companies, which was 50% in 1988, has since fallen steadily to 10% today. That caught the public's attention.

While the world's semiconductor market has continued to grow robustly at more than 5% per year during the past 30 years, Japan's market has remained flat. If the trend continues, Japan's market share would drop to almost 0%!? On the other hand, the global market is expected to grow at an even more rapid rate of 8% per year, riding on the tailwind of the digital revolution, and is on track to exceed $1 trillion by 2030, double the current amount.

Is there a turnaround scenario?

Simply put, the key to success in the semiconductor industry is aggressive investment in scaling technology.

However, it is difficult to make up for the time Japan lost in the last 30 years by conventional tactics alone. Therefore, it is also necessary to foresee the next battleground and make advanced investment accordingly ahead of the competition. This is referred to in the Japanese martial art of Kendo as "sen-sen no sen o utsu," or "anticipate your opponent's next move and strike preemptively" [11].

To make sense of the current complex environment, it is necessary to understand the three transformations shaping it.

© The Author(s), under exclusive license to Springer Nature Switzerland AG 2025   11
T. Kuroda, *The Super-Evolution of Semiconductors*, Synthesis Lectures on Engineering, Science, and Technology, https://doi.org/10.1007/978-3-031-60518-5_2

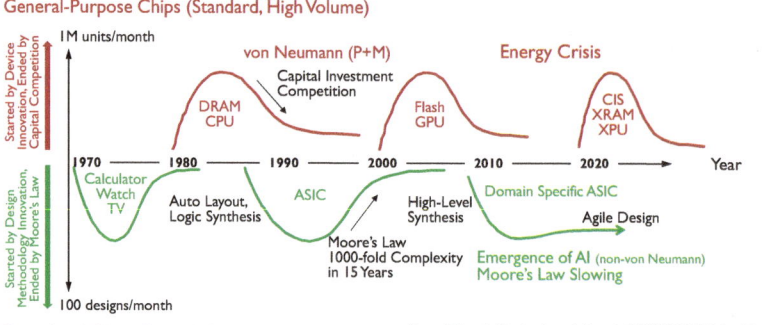

Fig. 2.1 The energy crisis created by the explosion of data and the slowing of Moore's Law is driving the resurgence of specialized chips

The first is the changing of the guard in the industry. The main battleground for logic semiconductors is shifting from general-purpose chips developed by chipmakers like Intel to specialized chips developed by chip users such as GAFA.

If you look at the deals that 25 of the leading US venture capital firms closed in the three years since 2017, you see substantial investments in specialized and AI chips amounting to nine times more than investments in memory.

It is a sign that a new era of specialized chips has arrived (Fig. 2.1).

The traditional approach in the semiconductor business is to mass-produce standardized, general-purpose chips. Still, there was a time in the past, from 1985 to 2000, when specialized chips were custom-made and produced in small quantities. Combining scattered logic from general-purpose chips into a single specialized chip could reduce the cost of manufacturing a product.

However, specialized chips were expensive to develop. As a result, computer-aided design (CAD) semiconductor technologies emerged in universities in the United States to overcome the challenge.

In spite of that, in 15 years, Moore's law increased the level of integration by three orders of magnitude, and eventually, design could no longer keep up. Thus, the era of specialized chips came to an end [12].

The looming energy crisis partly contributes to the current game-changing environment in the semiconductor industry. For example, analyzing exploding data with AI requires enormous amounts of energy. As a result, there is a need for specialized chips that consume orders of magnitude less power than general-purpose chips by eliminating unnecessary circuitry.

Specialized chips (hardware) accelerate AI processing, while general-purpose chips (running software) realize diverse functions. In other words, the proper division of labor between the two is essential for green growth.

2.1.1   Paradigm Shift

The second transformation shaping the current environment is a market wave.

A significant wave comes to the semiconductor market every quarter century. The next wave is now upon us.

Consumer electronics from 1970–1995, PCs from 1985–2010, and smartphones from 2000–2025. Japan caught the first wave but failed to capture the second and third. Therefore, it is crucial for us to prepare for the fourth wave.

Home appliances provide convenience in the physical space through analog technology. On the other hand, PCs create cyberspace through digital technology, while smartphones make cyberspace portable through wireless network technology.

The current fourth wave aims to solve economic development and social issues by tightly integrating cyberspace and physical space using sensors, AI, and motors. In other words, it creates a human-centric society using a digital twin, which Japan calls Society 5.0.

An example is robotics, where robots include cars, drones, and other mobile robots.

According to futurist Hans Moravec, the intelligence level of robots is currently at the level of mice. Still, it will evolve to the level of primates by 2030 and reach human-level intelligence by 2040. Intelligent robots will transform everything from mobility, logistics, and services to medicine, nursing care, and entertainment.

These are markets where Japan, a developed country facing many related challenges, can lead the world. In addition, they are areas where Japan can take advantage of its strength in the physical space, specifically, in the meticulous adjustment required for seamless system integration.

Of course, the fourth wave is not limited to this. It will test our ability to come up with new ideas, and at the same time, require agile development capabilities to immediately implement those ideas in a chip.

And the third transformation shaping the current environment is a paradigm shift in technology.

In the 1950s, computers were "wired-logic systems" programmed by rearranging the wiring between arithmetic units.

There are two drawbacks to this architecture. One is the "challenge of scale" where the size of the hardware prepared in advance limits the maximum size of the program. The other is the "challenge of wiring" where the number of connections becomes enormous as the size of the system grows.

To work around these challenges, mathematician John von Neumann invented the "stored-program architecture," also known as the "von Neumann architecture." In this architecture, memory is used to store data and instructions for data movement and computation, and a processor is used to interpret these instructions sequentially to perform computational tasks. This revolutionary change of architecture aimed to solve the problem of scale constraints by having a single processor execute a different instruction in each

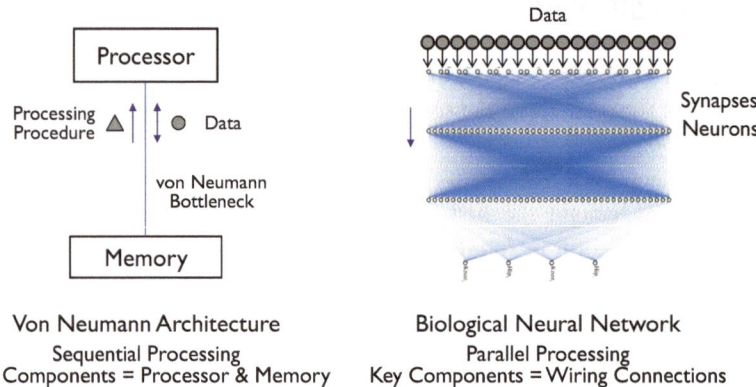

Fig. 2.2 From von Neumann to neural networks

processing cycle rather than preparing multiple arithmetic units and physically connecting them to implement the specific computation.

On the other hand, the integrated circuit (IC), invented in 1958 by electronic engineer Jack Kilby, was born out of the study of the "challenge of wiring" from various angles. Jack Kilby successfully solved this problem by using photolithography to integrate multiple transistors on a single chip and wire them together all at once.

Two paradigm shifts are now taking place in the field of computer architecture, which has been defined by these two architectures for over half a century.

One is the shift from von Neumann architecture to neural networks (Fig. 2.2).

Instead of repeatedly and sequentially processing data that is moved back and forth between the processor and memory, neural networks process data simultaneously in parallel as it flows through the network. The result is a major improvement in energy efficiency.

The use of the von Neumann computer architecture resulted in mass consumption of processors and memory chips. But future growth is anticipated for specialized chips which implement neural networks for AI processing. The core hardware for the industry will change from processors and memory chips to the physical wiring of neural networks. This is just like the evolution in living organisms from the brain stem and cerebellum to the cerebrum.

At birth, our brain has only about 50 trillion synapse connections. But the number continues to grow and increases by 20-fold by the time we enter elementary school. After that, as we learn, unused connections are gradually removed, resulting in a highly efficient brain network without waste at maturity. In other words, our brain is incomplete at birth, grows through playing, and achieves high efficiency through learning.

The formation of neural networks follows a similar path. There is currently a lot of active research on pruning of neural networks through machine learning.

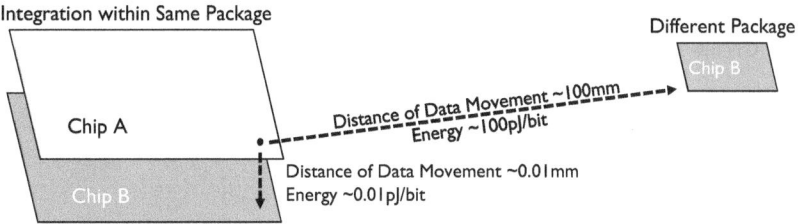

3D integration reduces energy required for moving data by orders of magnitude.

Fig. 2.3 From scaling to 3D integration

The second paradigm shift moves the industry from process scaling to 3D integration. (Fig. 2.3).

Process scaling is finally approaching its limits. 3D integration can reduce the energy consumed in moving data by orders of magnitude. It is like taking the data that you previously had to go to the National Diet Library to retrieve and placing it within your arm's reach.

As a result of these paradigm shifts, we must once again overcome the two fundamental challenges we faced in the 1950s. It presents an opportunity for disruptive technologies to shine as we approach the end of Moore's Law.

Green Growth Strategy

From our analysis so far, we can see that the energy consumption problem is the root cause of various transformations in the industry.

In order to increase energy efficiency, the industry is going through a changing of the guard from general-purpose to specialized chips, computer architecture is shifting from von Neumann to neural networks, and the focus of integration technology is moving from scaling to 3D integration.

Meanwhile, society is evolving from the capital-intensive industrial society to a knowledge-intensive intelligent society. Value is no longer found in low-cost chips that integrate a massive number of transistors. Instead, value is being shifted to the ability to process large amounts of data in a highly energy-efficient manner and the superior services that such ability creates.

However, going forward, the move towards carbon neutrality will create restrictions that weigh heavily on the industry. We must cut energy consumption aggressively. **We will certainly have to transform our growth strategy from the current greedy strategy to a green strategy** (Fig. 2.4).

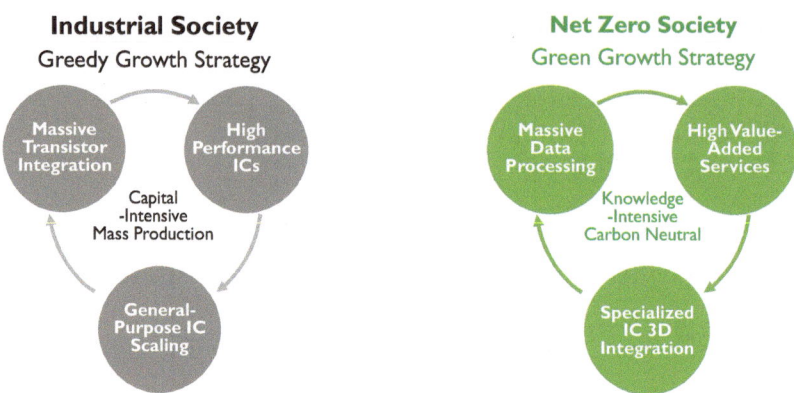

Fig. 2.4 Transition from greedy to green growth strategy

The "three arrows" of Japan's green growth strategy are the creation of 3D integration technology to eliminate the bottleneck in data processing, the construction of a platform for agile development of specialized chips, and the preservation of the domestic ecosystem.

Without improvement in energy efficiency there will be no growth; without improvement in development efficiency there will be no specialized chips. In other words, the highest-priority challenge going forward is the pursuit of time performance. Since time is money, the conventional cost performance is also factored into time performance.

Former British Prime Minister Winston Churchill, in addressing the challenges faced by his country, said, "One ought never to turn one's back on a threatened danger and try to run away from it. If you do that, you will double the danger. But if you meet it promptly and without flinching, you will reduce the danger by half."

Meanwhile, Intel co-founder Robert Noyce once said, "Optimism is an essential ingredient for innovation. How else can the individual welcome change over security, adventure over staying in safe place? [13]".

With resolve and optimism, let's turn Japan's semiconductor industry around, starting today.

2.2   From General-Purpose to Specialized Chips: Game Changer in the Semiconductor Industry

The Eras of General-Purpose and Specialized Chips

General-purpose products by nature are used in many different applications. Consequently, they are produced in large quantities, resulting in low cost, which in turn leads to wide adoption. Meanwhile, specialized products, although expensive, offer better performance, quality, and reliability.

In the semiconductor industry general-purpose chips dominate. With an approximate annual revenue of $500 billion and production volume of 2 trillion chips, the average unit price is only around 25 cents.

Even the most advanced chips made in state-of-the-art factories, which cost tens of billions of dollars to build, sell for only a few dollars. It is therefore a low-margin business which relies on volume for profit.

Thanks to the adoption of the von Neumann computer architecture, the industry has been able to generate demand for large volumes of general-purpose chips.

Under the von Neumann architecture, a processor reads the processing procedure and data from memory, processes the data following the processing procedure, and writes the results back to memory. By looping through this process repeatedly in sequence, it is possible to implement processing of any complexity. Meanwhile, by changing the processing procedure (or program), it is possible to perform any kind of processing.

By adopting this architecture, the evolution of the computer has followed a scenario where general-purpose hardware—processors and memory—are produced in large volumes to drive adoption, and software is used to tailor the hardware for various applications [14]. As a result, the focus of the semiconductor business has been on the production of large volumes of processors and memory at low costs. More recently, the rise of big data has resulted in sensors being added as another mass-produced device type.

Competition in such a business is in terms of capital investment. When the business potential of a newly invented device such as DRAM, flash memory, CPU, or GPU is recognized, large capital investments follow. This quickly leads to fierce competition, resulting in industry realignment and eventually consolidation.

Japanese companies won the competition in device innovation but lost the competition in capital investment [15].

On the other hand, specialized chips had their share of success too—ASICs (application-specific integrated circuits) had a sizable market between 1985 and 2000.

Glue logic which interconnects processors and memory varies from system to system. While systems were previously implemented using a combination of standard logic chips, such chips were later integrated into ASIC to reduce system cost and area.

Another important reason that turned ASIC into a profitable business was the adoption of computer-aided design (CAD) to dramatically reduce both development cost and time. Using CAD, a complicated chip that previously would take 100 engineers a year to design could be designed by one engineer in one month.

In the 1980s, research and development efforts led by the University of California at Berkeley resulted in the creation of automatic layout and logic synthesis technology as well as the birth of chip design tools vendors. Furthermore, a semi-custom manufacturing process was developed where a semi-finished chip was made first like a standard product, which was then tailored towards different applications by customizing the interconnect layers.

Using these design methodology innovations, chip development productivity was improved by three orders of magnitude in total.

Nevertheless, with Moore's Law increasing integration density by three orders of magnitude in 15 years, even with computer-aided design, it took more man-hours than ever to develop specialized chips. This contributed to profit erosion of the ASIC business, which eventually resulted in its demise.

In this manner, **an era of general-purpose chips is started by device innovations and ended after fierce competition in capital investment. Meanwhile, an era of specialized chips is started by design methodology innovations and ended by Moore's Law** [16].

Game Changer: GAFA's In-house Development of Specialized Chips

However, we are now in the middle of a game changing trend—IT giants such as GAFA have embarked on in-house development of specialized chips, since they find it difficult to compete by relying on general-purpose chips procured from specialized chipmakers the likes of Intel and Qualcomm.

There are three reasons for that.

The first reason is **the unique energy crisis faced by data companies**. The explosive growth of data and the rising sophistication of AI processing have fueled the energy crisis, as described earlier.

Without further advance in low-power technology, by 2030, IT machines alone would consume about twice the world's total power output of today, increasing to two hundred times by 2050.

If digital transformation consumes so much energy as to destroy the environment, it will make a sustainable future impossible.

At the beginning, a chip consumed just about 0.1 W of power. Under the ideal scaling scenario, its price performance could be improved while its power density remains constant.

In reality, however, as a result of prioritizing price performance over power, power was allowed to increase to achieve price performance improvement beyond what was possible under the ideal scaling scenario, resulting in a 1000-fold increase in 15 years and reaching 100 W in 2000. Chip power density is now more than 30 times that of a hotplate used for cooking. As a result, it requires a tremendous amount of power to cool a cloud server.

When chip power exceeds the cooling limits, even if we can increase integration density, we cannot power up and use all the transistors at the same time. The more chip power exceeds the cooling limits, the more there are unused transistors. For example, while unused transistors account for about 75% of the total in the 7 nm generation, the ratio is expected to grow to 80% in the 5 nm generation.

Under such constraints, only those who can improve energy efficiency tenfold can achieve a tenfold increase in computing performance, or a tenfold increase in smartphone battery life [17].

The second reason for the shift to specialized chips is the rise of AI. AI in the form of neural networks and deep learning offers a new way of information processing to owners of data.

Similar to our brain, neural networks are based on wired logic where functionality is defined by how the components are wired together. Furthermore, data is processed in a parallel fashion as it flows through the network. Since parallel processing enables lower operating frequency and hence lower chip voltage, wired logic can improve energy efficiency by more than tenfold compared to von Neumann architecture where data is processed sequentially.

The third reason for the shift to specialized chips is the adoption of the fabless model by the semiconductor industry. Under this model, pure-play foundries such as TSMC offer manufacturing services to the world, which enables any user adopting a business model of providing superior AI performance to develop their own chips.

For companies offering a hardware platform solution that drives the demand for sufficiently large chip volumes, the fabless model allows them to design specialized chips to more quickly realize chips with higher performance and at lower cost than procuring from specialized chipmakers.

Considering Manufacturing in a Knowledge-Intensive Society

As Alan Kay once pointed out, "People who are really serious about software should make their own hardware." In system development, it is necessary to think both hardware and software.

The choice of architecture depends on the type of processing performed. Logical and arithmetic processing which requires versatile controllability is better performed using the conventional solution of general-purpose chips implementing the von Neumann architecture. Meanwhile, intuitive and spatial information processing which requires sophisticated AI computation is better performed using neural networks implemented in specialized chips, which, for reasons previously explained, achieve high energy efficiency. With the shift of chip application from products to services, the quest for a new, matching architecture has commenced.

The fact remains that the choice between general-purpose and specialized chips involves tradeoffs between low cost and high performance.

For illustration, let's look at data communication. Communication infrastructure which does not drive large volumes can be created by adopting general-purpose hardware and implementing unique functionalities using virtualization technology. However, at the edge where there are comparatively large device volumes, specialized chips can be utilized to boost performance to enable distributed processing local to where data is generated.

The development of specialized chips is knowledge-intensive, not capital-intensive. Development of automatic layout and logic synthesis was previously driven by the University of California at Berkeley. Similarly, this time around, **university research will play an integral role in creating the fundamental knowledge required for specialized chips development, including knowledge for automatic generation of functionalities and systems** [18].

The twentieth century was the century of "general-purpose." After the war, in a quest for material gratification and economic efficiency, economic growth was driven by mass production of standardized products.

However, as modern society matures, our emphasis has shifted from collective growth to personal fulfillment. This has resulted in the transition from an industrial to a knowledge-based society.

While this transition was spreading from developed to developing countries, Japan was able to enjoy a period of prosperity by continuing to mass-produce standardized products. But eventually when the transition was complete, Japan fell behind other Asian nations due to its focus on manufacturing and hence slow adaptation to the transition.

The twenty-first century looks to be the century of "specialization." The center of our value is shifting from being capital-intensive to being knowledge-intensive, from scale to knowledge, from increase in quantity to increase in quality, from material to spirit, from convenience to joy, from products to services, from large volume to large variety, from standardization to individualization, from what everyone can do to what no one else can do.

How should the manufacturing industry adapt to such a shift? It is d.lab's mission to search for an answer.

2.3   From Staple Food of Industry to Brain Cell of Society: Semiconductors in the Post-COVID Era

The Energy-Consuming Remote Society

An American friend of mine built his home in the middle of a forest and works from there remotely. Since he is an EDA developer, he can work from anywhere as long as he has a computer and internet access. So I thought. However, …

COVID-19 has swung open the door to a remote society. Online meetings work better than we thought and are great for discussions involving 3 or more people.

Even international conferences with as many as 3000 participants have gone online.

Back in 2005, in my opening speech as program chair at the reception of an international symposium held in Kyoto, Japan, I said the following:

Imagine this. In the future, we may hold an international conference on the internet. Everything is done online – research paper presentations, panel discussions, even hallway conversations. And everyone can participate from home. 'What about the banquet?' You ask. We get pizza from delivery and beer from the fridge ... That doesn't sound too exciting, does it? Instead, let's enjoy tonight Kyoto cuisine and Japanese sake while exchanging thoughts with old friends. Cheers!

But now organizers of international conferences are probably worried that everything in Pandora's box is out, since even drinking parties have been taken online.

The surge of big data and the advance in AI processing are fueling a digital transformation and enabling data-driven services, which in turn are leading to explosive growth of energy consumption in society.

As noted above, **it is forecast that by 2030 IT-related equipment alone will consume close to twice the total electrical power generated today. And by 2050, that power consumption will further increase to 200 times of today's total power output.**

One of the reasons is the exponential growth of data communication. While total annual IP traffic in 2016 was 4.7 ZB, it is expected to grow 4 times to 17 ZB by 2030 and 4000 times to 20,200 ZB by 2050.

Local production and consumption of information is essential from the energy efficiency perspective. On the other hand, the accumulation and monopolization of information by giant IT companies will continue.

Adding to that is the increase in sophistication of AI processing. It requires a tremendous amount of computation to extract meaning hidden in data to drive services that deliver value to society.

In fact, since the advent of deep learning, the computational complexity of AI processing has increased by four orders of magnitude over the past decade. Meanwhile, the power efficiency of general-purpose processors has improved by only one order of magnitude.

In other words, sustainable growth of society will not be possible without improving the energy efficiency of telecommunications equipment and computers by orders of magnitude.

It is semiconductors which are causing the rapid growth in energy consumption. It is also semiconductors which hold the key to solving the problem.

From Being the Staple Food of Industry to the Brain Cell of Society

In 2019 the world made a total of 1.9 trillion semiconductor chips.

The market breakdown by sector is as follows: manufacturing 15%, healthcare 15%, insurance 11%, banking and securities 10%, wholesale and retail 8%, computers 8%, government 7%, transportation 6%, public utilities 5%, real estate and business services 4%, agriculture 4%, communication 3%, others 4%.

Indeed, semiconductors are used in every corner of society. But, on the other hand, you may be surprised to learn that the communication market is still small.

However, as mentioned earlier, the volume of communications is expected to explode in the near future. It is Post-5G, the next-generation communication technology, which is expected to drive the demand for next-generation semiconductors.

Post-5G uses high frequency bands. The higher the frequency, the more the signal travels in a straight line, but the shorter the distance it can travel. Consequently, higher frequency requires more base stations.

In addition, since low-latency, high-value services are desired, base stations are expected to perform sophisticated data processing. That is why Post-5G is anticipated to drive the demand for next-generation semiconductors.

Going forward, services are expected to create a large semiconductor market in addition to IoT, digital medical and healthcare including remote medical services, and mobility. Together they will form the nervous system of society.

In other words, **semiconductors are evolving from being the staple food of industry to the brain cell of society, making them more and more a global commons—a resource shared by the world**.

As semiconductors transform from being a component in the industrial society to a strategic resource that supports an intelligent society, the value proposition of semiconductors will shift from cost to performance, especially power performance. In addition, since they will be used in infrastructure, time-to-market and reliability will also become important.

The only way to solve society's energy problem is to increase semiconductors' energy efficiency. Compared to general-purpose chips, specialized chips can achieve a power efficiency that is two orders-of-magnitude better [19]. That is because specialized chips created for well-defined users and usage scenarios have no superfluous circuits which are required in general-purpose chips to respond to diverse demands from unspecified users and to future-proof the product.

However, since the development cost of specialized chips is high, not everyone can afford to develop them.

As a result, the environment for specialized chips development in Japan has become unfavorable, and a hollowing out of the market has begun.

Therefore, **it is essential to reduce the development cost of specialized chips by an order of magnitude so as to enable anyone with a system idea to design a specialized chip, as well as reduce energy consumption by an order of magnitude using cutting-edge semiconductor technology. This is necessary for realizing a data-driven society**.

To enable semiconductors to shift from being the staple food of industry to the brain cell of society, it is also necessary to transform the structure of the industry from the capital-intensive industry of the last half century to a knowledge-intensive industry for the next half century.

To Create a Digital Civilization

Yuval Noah Harari, the author of Sapiens: A Brief History of Humankind, recently wrote an article where he warned that technology can become a spy who carries out "under-the-skin surveillance" by monitoring our biological conditions.

In combating the spread of COVID-19, surveillance has become more widespread in society. The impact of technology on society has become exceedingly large. What is that going to do to our civilization? Some even believe that we have come to a critical juncture.

Technology can implement any human intelligence. Therefore, while semiconductors can threaten our security and privacy, they can also be the solution.

However, it will naturally increase energy consumption in order for semiconductors to offer sophisticated security and privacy protection. Therefore, once again it brings us back to semiconductors' energy problem.

Further ahead, there is also the problem of the mind.

Digital excels at handling logic, while analog emotions. And now we are starting to use digital to pursue happiness.

We cannot blindly promote the connection of our brain to the internet without first laying the necessary foundation in place which includes sensors and actuators that convert our five senses to digital and vice versa, control technology that feeds back our senses, engineering of technology for value transactions (such as blockchain), and a legal system that prevents technology from making society dangerous,

A long, long time ago, the brain created society and gave birth to the mind. Humans came to recognize their intentions and created languages to convey them, while also developing mathematics to expand their cognitive ability through logical thinking. Eventually, **mathematics** evolved into a system of abstract symbols that exceeded our subjective intuition and **overflowed from our brain to create the computer. The computer then gave birth to chips, which in turn enabled downsizing of the computer as a result of exponential growth through scaling. At some point the computer will become so tiny that it can be put back into our body** (Fig. 2.5).

Fig. 2.5 Chip power efficiency improved by 3 orders of magnitude in 20 years and expected to approach that of the brain by 2030

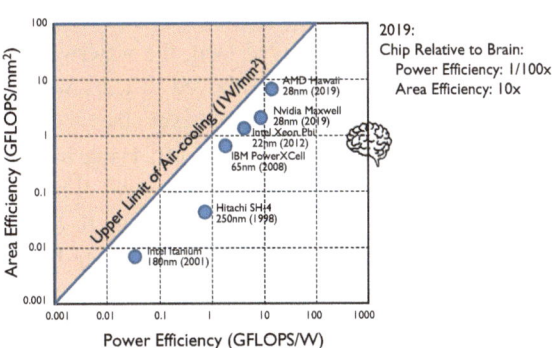

2.4   From Dams to Semiconductors: Infrastructure of a Digital Society

Hatta Dam and TSMC

Wushantou Dam in Taiwan was constructed 100 years ago. Up until Hoover Dam in the United States was completed, it was the largest dam in the world. The dam was built under the supervision of Japanese civil engineer Yoichi Hatta. In honor of his contribution, the dam is also known as Hatta Dam.

Mr. Hatta studied Civil Engineering at the Tokyo Imperial University, the predecessor of the University of Tokyo. After graduating in 1910, he joined the Civil Engineering Division of the Taiwan Governor-General Office as an engineer. After surveying the Chianan Plain, a wasteland extending across Southern Taiwan, he realized that the farmers in the region constantly suffered from drought, heavy rain, and poor drainage because of an inadequate irrigation system. In response, he proposed irrigation system development works to convert the wilderness into a grain-producing region and obtained approval from the Japanese government. The works were executed by a union of the beneficiaries, with half of the costs covered by Japan's national budget. Mr. Hatta voluntarily gave up his status as a government employee to become an engineer for the union to spearhead the dam construction.

The project cost a total of 54 million yen. At the time, it was Japan's biggest construction project ever. The construction of the dam, which involved tunneling through the 3,078-m-high Wushan Ridge to divert water from the main Zengwen River, resulted in many casualties. And until the completion of this grand project, Hatta lived with his wife and eight children in a crude Japanese-style house with about 100 m^2 of living space built on the construction site.

It took 10 years to complete the dam.

The dam is complemented with a waterway that is 16,000 km (about 10,000 miles) long and intricately woven across the Chianan Plain. It is the Great Wall of Water, so to speak. Since the Great Wall of China is only 2700 km long, the waterway far exceeds its length.

When water first emerged from the waterway, the 600,000 local farmers hailed it as "water from God" and were moved to tears with joy and gratitude. The Chianan Plain has continued to be well irrigated ever since.

In Chianan a bronze statue of Mr. Hatta has remained to this day. It was carefully preserved by the locals even during the war. It has become a common practice for passing local farmers to stop in front of the statue to quietly pay respect by putting their hands together. In addition, on the anniversary of Mr. Hatta's death, the locals will visit the tomb of him and his wife for a memorial service.

Mr. Hatta was a Japanese who loved Taiwan. And as the Father of the Chianan Irrigation System, he became a Japanese loved by Taiwan.

Now, 100 years after the construction of Hatta's dam, Japan and Taiwan are joining forces again to embark on a major, historic undertaking.

This time around, it is semiconductors instead of a dam.

Soil is replaced by high-purity silicon [20].

Water is replaced by data, and water usage by data utilization.

Society is evolving from an agrarian society, Society 2.0, to a human-centric society, Society 5.0, and the Great Wall of Water will be replaced by the Great Wall of Data.
A small chip has numerous wires integrated into it, through which data travels. When all of these wires are pulled out from the chip and connected together, they can be as long as 10 km.

TSMC's factories manufacture millions of wafers every month, each containing about 1,000 chips. If the wires from all these chips were connected together, the total length could reach 1 billion kilometers, allowing data to travel around the Earth 25,000 times.

Kikuyo-cho, Kumamoto Prefecture, Japan—Many construction cranes stand side by side on a vast site of over 20 hectares, where TSMC's Kumamoto plant is being built. The new plant will employ 1,700 workers. Of these, 300 will come from TSMC, 200 from Sony Group companies, and the remainder will be newly hired. Competition for local talent has begun, which has pushed up wages for engineers.

Logic chips at 28 and 22 nm will be manufactured at this plant. These are the volume zone semiconductors most needed and produced in Japan. In the future, 16 nm and 12 nm FinFETs will be made as demand shifts to finer generations. Japan was not able to continue investment in semiconductors beyond 28 nm. Therefore, we will rejoice the arrival of "data from God" when production begins.

In addition, TSMC has opened a 3DIC R&D center in Tsukuba City, Ibaraki Prefecture. Chips will be stacked in 3D to shorten the distance of data movement. This will require Japan's superior materials. We aim to discover new materials and work with the 3DIC R&D Center to find a way to make the most out of Japan's materials.

The Infrastructure of a Digital Society

Infrastructure of the twentieth century includes the transportation infrastructure of roads, ports, railroads, and airports, the urban infrastructure of water and sewage systems, and the energy infrastructure of power generation and transmission. The infrastructure of the twenty-first century will consist of semiconductors and the advanced computing and communications networks that use them.

In its postwar reconstruction, Japan achieved success as a capital-intensive industrial society and industrial power. Eventually, people became aware of the limits of growth in mass production and mass consumption; we came to believe that the society we should aim for in the future is a human-centric society that rises above the industrial and information society. It is a knowledge-intensive society in which people can share their knowledge and wisdom while utilizing data. In other words, it is a shift from a capital-intensive industrial society to a knowledge-intensive, knowledge-value society.

As society undergoes a paradigm shift from a capital-intensive to a knowledge-intensive society, the industrial structure will also change. In a capital-intensive society, value is created by goods. Materials are the resource, which are combined to make parts, and parts are assembled to make products. Semiconductor chips are parts, and knowledge and information embedded in designs and services provided with products create user satisfaction and hence value.

However, when society changes to a knowledge-intensive society, the roles in value creation are reversed.

Value shifts from objects to knowledge and information. Materials are replaced with data as the resource: IoTs collect data which is then analyzed by AI. The results are provided to users in the form of tailored services and solutions. The semiconductors that deliver them will create user satisfaction and hence value by making their device's battery last longer or data processing faster.

Thus, instead of roads, ports, railroads, and airports used to transport materials, IoTs, 5G, and AI used to move data will become the infrastructure of the digital society.

For semiconductors as a component business, cost is the top priority. When the parts are standardized and similar, cheaper is better. However, as scaling becomes more and more difficult, competition also began to be based on performance. **A commonly used metric these days is PPAC (power, performance, area, and cost). In other words, it is cost performance that matters**.

Time performance will be required for semiconductors that support the infrastructure of the digital society. This is because replacement demand in the infrastructure market is small, and products introduced first in the market will be used for a long time. **PPACT, to which time-to-market is added, will therefore serve as the new metric**.

Japan's infrastructure in the twentieth century was excellent: roads extended everywhere, and transportation systems operated on time. The superior infrastructure of the twenty-first century will consist of high-speed wireless networks that enable utilization of advanced computing resources from anywhere in the country.

In order to survive the Great Depression of 1929, the U.S. built infrastructure through public works projects such as the construction of multi-purpose dams under the New Deal. Semiconductors are a foundational technology that supports the infrastructure of the digital society. Therefore, I think our society is now in need of a Digital New Deal.

The Teachings of Professor Hiroi

The time was once again 100 years ago. Professor Isami Hiroi was teaching Civil Engineering at the former Tokyo Imperial University. In response to the question of what engineering is for, he said the following.

Engineering will have no meaning if all it does is to make the life of humankind more complicated. The only way we can find meaning in engineering is if we can use it to shorten time from taking days to taking hours and complete in an hour work that would otherwise require a day, so that we can use the time saved to quietly think about our life, to reflect, and to be close to God.

Mr. Hatta must have learned from Professor Hiroi.

Today we must once again recall the spirit of Mr. Hatta and the teachings of Professor Hiroi. The commencement of the historic undertaking in Kumamoto and Tsukuba that crosses ethnic and national boundaries between Japan and Taiwan is filling me with great expectations and awe and giving me hope and courage.

2.5   [Column] Rapidus' Strategy

TSMC's lineup offers everything.

As of this writing, their logic chips start with the most advanced 3 nm (nanometer) process and go on to the conventional processes of 5 nm, 7 nm, 10 nm, 16 nm, 20 nm, 22 nm, 28 nm, 40 nm, 65 nm, 90 nm, 0.13 μm (micrometer), 0.18 μm, 0.25 μm, 0.35 μm and 0.5 μm. In fact, TSMC now offers all process technologies from since the 1980s, spanning 16 generations.

There is also a wide variety of product offerings. In addition to logic chips, TSMC offers analog chips, high-frequency wireless chips, mixed DRAM, mixed non-volatile memory, image sensors, high-voltage devices, power devices, and MEMS (micro-electromechanical systems). For each of them, multiple generations of process technology are offered.

The company boasts a total lineup of 80 product types.

In addition, the production volume is huge. Wafer production capacity is 12 million wafers per year (300 mm equivalent). This accounts for 60% of global foundry production. In response to strong demand, TSMC has announced plans to significantly increase its production capacity over the next several years. The new plant being built in Kumamoto Prefecture aims to produce 540,000 wafers per year.

The strategy of Japan's new foundry company, Rapidus, is in stark contrast.

Rapidus will produce only the most advanced products in small quantities with a short turnaround time, starting with the 2 nm process, and always be using only the three most advanced generations of process technology.

Objections to this strategy are fierce.

"Why 2 nm all of a sudden?" "It's reckless. Since we've been stagnant for 20 years, we should start with smaller steps."

"Will it be profitable at the cutting-edge process?" "Will there be users?"

For me, I think Rapidus has the right strategy.

First, cutting-edge processes are profitable.

In fact, TSMC's main source of income is from the cutting-edge technologies: 5, 7, and 10 nm together account for the majority of its total sales.

This seems to defy common sense, as noted by Rapidus President Atsuyoshi Koike, who prefaced his presentation of this data by saying, "This is against conventional wisdom."

Until now, semiconductors have been subject to fierce price competition. If a company were to pass on all the costs of the expensive, cutting-edge manufacturing equipment to prices, it would lose out to the competition. Therefore, it has been believed that semiconductor companies should compete on price, even for the state-of-the-art technology. Only after the equipment installed in new plants has been fully depreciated can it finally generate profits.

Semiconductor manufacturers generally depreciate their equipment over a period of three to five years. Thus, the conventional understanding has been that cutting-edge technology is not profitable, but it will generate profits one or two generations later.

However, the competitive environment has changed. The number of players in the most advanced technology markets has been decreasing yearly.

For example, at present, only TSMC can mass-produce 3 nm, Samsung Electronics and Intel can fabricate 5 nm, seven manufacturers can fabricate 7 nm, and nine companies can fabricate 22 nm. In other words, the cutting-edge is a de facto oligopoly market.

And there is always demand for cutting-edge chips.

Similar doubts have emerged many times in the past with memory chips. "Who could possibly use up that much memory?" The same question is always asked, but there are always those who plan for the next generation of memory.

As expectations for a data-driven society rise, demand for servers for data centers will also increase. And the era of the great DRAM competition is expected to arrive: the market forecasts that DRAM production will surge to 170 billion chips (2-gigabit equivalent) in 2025, 2.5 times as many as in 2020.

Memory chips and digital are the twins spawned by the von Neumann architecture. If the memory chip market grows, so will the digital market.

In short, the demand for cutting-edge products is growing.

The demand is also spurred by the fact that next-generation chips with 30% higher performance may no longer be available after a two-year wait, as was the case in the past.

Under these conditions, where demand is growing and supply is scarce, wouldn't the supplier have leverage over price? The fact that more than half of TSMC's revenue is attributed to cutting-edge semiconductors implies that.

If a manufacturer can adequately pass the cost onto customers without losing out to competitors, it can profit from being a first mover.

Let's also consider this from the viewpoint of technology. For example, I have heard criticisms such as, "Instead of jumping straight into 2 nm, wouldn't it be more reasonable to start with 5 nm, hone one's skills at 3 nm, before trying 2 nm?".

In that case, there are two risks. First, TSMC, Samsung, and Intel have significant cost advantage over any new challenger, having already completed depreciation of their manufacturing equipment at 5 nm. It would be easy for them to block Rapidus from entering the market if they wanted to.

Furthermore, even if Rapidus perfects its skills with FinFET technology at 5 and 3 nm, the learning may not be transferrable to 2 nm, which will use GAA, a radically different transistor structure.

Therefore, the strategy of re-entering the market at 2 nm, where competition will start afresh, and taking on small-volume production, which the mega foundries do not cater to, rather than competing with the mega foundries head-on, would be welcome by the market.

Innovation starts with low-volume production. The key to innovation is getting to market quickly. Therefore, rapidity creates value. That is the strategy of Rapidus.

Rapidus does not compete with mega foundries; rather, it complements them. Rapidus does not offer everything like they do, but only high-end products. Rapidus will not mass-produce, but will produce with a short turnaround time. Eventually, products that require mass production will emerge from this process. In this way, Rapidus will contribute to society in a complementary manner and coexist with mega foundries. And most importantly, there should be users who need high-end products made in a short turnaround time.

If we were to compare the semiconductor scaling race to a marathon, we would say that the race was nearing its end, and the top runners were starting to enter the final spurt, while others were gradually dropping out of the lead pack. This was no longer the time to try to gain an advantage over others around you, by reducing air resistance or monitoring your rivals. If God were to give you strength at this moment, you should aim at the backs of the leaders and run as fast as you could. You should aim for a decisive comeback.

This reminds me of a famous quote from John Maynard Keynes.

"The difficulty lies, not in the new ideas, but in escaping the old ones, which ramify, for those brought up as most of us have been, into every corner of our minds." (*The General Theory of Employment, Interest and Money* by John Maynard Keynes. Macmillan, 1936).

Structural Transformation: More Moore

3

3.1 A Short History of the Brain, Computer, and Integrated Circuit, and One Scenario of the Future

The Birth of the Brain, Computer, and Integrated Circuit

Some 13.8 billion years ago, a gigantic energy mass suddenly came into being. That is the basis of the Big Bang theory.

Energy and matter interacted ($E = mc^2$), and the universe rapidly expanded. What started as a little tremor created the Galaxy, and the Earth was born 4.6 billion years ago.

While matter transformed following physical laws, life came into being 4 billion years ago, which replicates itself by coding and storing its structure in the form of DNA.

Life, using mutation and survival of the fittest as strategy, survived in an uncertain environment and diversified by evolving from single-celled to multicellular organisms, to plants and animals.

As animals continued to evolve, eventually they developed the brain which is the central nervous system that determines their action based on information they collect from the outside world. Then 7 million years ago, the human brain further advanced, differentiating human beings from other mammals.

In order to survive, human beings learned the importance of working together. In other words, it is the brain that created society and gave birth to the mind. Furthermore, we developed languages to communicate our intentions, in addition to acquiring the ability to think logically.

Mathematics was born 3000 years ago.

Mathematics expanded our cognitive capacity. The four great ancient civilizations used computing tools and principles such as Pythagoras' theorem for tasks like calculating taxes and surveying land. Later in the fifth century B.C. in ancient Greece, rather than its use for computation, the inside world of mathematics became the subject of academic pursuit, and mathematics evolved from being a tool into a way of thinking.

With developments such as the advance of algebra in Arabia in the 7th Century and the invention of symbolic algebra during the Renaissance in the 15th Century, mathematics spread without constraints and became ubiquitous. Then in the 17th Century, the development of calculus enabled the inquiry into the world of infinite. As a result of close examination of concepts such as limits and continuity, abstract symbolism was born which reaches beyond subjective intuition.

By the 20th Century, efforts were started to make mathematics about how to perform mathematics. By completely shedding ambiguous ideas such as physical intuition and subjective feeling, **the mathematics that flowed out of our brain gave birth to the computer as machines that perform computation.**

Early computers were wired-logic machines, where programming was achieved by changing the wiring between computational units, as explained in Chapter 2.

Also discussed in Chapter 2, this architecture faced two challenges. The first was the challenge of scale where the scale of problems that could be solved was limited by the scale of the hardware. The second was the challenge of wiring where the number of interconnections exploded as the scale of the system rose. These problems were solved by von Neumann and Jack Kilby.

By integrating and parallelizing simplified and miniaturized computing resources on a chip to create an integrated circuit (IC), computer performance has improved dramatically. High-performance computers in turn enable the design of even larger-scale integrated circuits. Following Moore's Law, which is a rule of thumb driving IC scaling, computers and integrated circuits have both advanced tremendously (Fig. 3.1).

Growth and Limitations of Integrated Circuits

The cost performance of the IC can be exponentially improved by process scaling.

Cost is determined by lithography. But as lithography technology approaches the limits of scaling, complex processes, such as combining multiple photomasks in double and triple patterning, are employed to achieve further miniaturization. As a result, costs

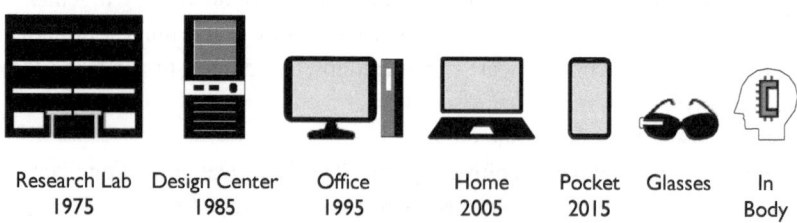

| Research Lab | Design Center | Office | Home | Pocket | Glasses | In |
| 1975 | 1985 | 1995 | 2005 | 2015 | | Body |

Fig. 3.1 Chip scaling enables computer downsizing, which in turn enables further scaling, resulting in the two advancing in tandem

increase, and transistor prices rise. Specifically, cost per transistor started to rise in 2015 in the 16 nm generation.

However, with the introduction of extreme ultraviolet (EUV) lithography in 2019 in the 7 nm generation, transistor unit cost is expected to fall once again. This is because the process is simplified, and manufacturing costs are reduced.

Therefore, the recent issue of IC lies not in cost but in the limits of performance improvement. The most pressing issue is that power, or heat generation, has reached its upper limit. As a result, no matter how much circuitry is integrated, performance cannot be improved any further.

It is computational performance per unit power consumed, or power efficiency (GFLOPS/W), which controls the destiny of Moore's Law. **In other words, performance cannot be improved without improving power efficiency.**

Power has been increasing as a consequence of scaling. Since transistors operate under electric field effect, if device is scaled while keeping electric field constant, power should remain constant.

But in reality, between 1980 and 90s, device was scaled without proportional reduction in supply voltage in order to achieve additional circuit speedup. Consequently, power increased fourfold every 3 years, resulting in a 3 orders-of-magnitude increase in 15 years.

Although voltage finally started to drop in 1995 when power became exceedingly high, electric field inside the device has become so strong that current is not decreasing adequately. As a result, power has continued to double every 6 years.

Since the power increase is a consequence of scaling, it is not easy to resolve. It requires thinking from first principles.

There are three ways to reduce power – lowering voltage V, capacitance C, and the frequency of switching $f\alpha$ [21].

While lowering voltage is effective in reducing power, it has its limits, imposed by leakage.

Due to quantum effect, current leaks through the insulating gate oxide. Consequently, there is a limit on how much the oxide layer can be thinned. Scaling the transistor without thinning the oxide layer accordingly results in the gate not being able to completely turn off the transistor.

As a result, further reduction in voltage leads to increase in leakage, making leakage current the dominant component that increases overall power. The maximum power efficiency of today's processors is achieved at an optimal supply voltage of around 0.45V.

To reduce leakage, materials, processes, and structures have been modified. For instance, gate control can be strengthened by wrapping the gate around a transistor built in 3D. Such a structure in the form of FinFET has achieved better than expected leakage reduction in the 7 nm generation.

From General- to Special-Purpose, 2D to 3D

At room temperature, the theoretical operating voltage limit for cascaded CMOS gates is 0.036V. In other words, there is room for another order-of-magnitude reduction in voltage, which translates into another 2 orders-of-magnitude reduction in power.

Another way to improve power efficiency is to reduce capacitance. **Compared to general-purpose chips such as CPU and GPU, specialized chips like ASIC (application-specific integrated circuit) and SoC (system-on-chip) can achieve tenfold improvement in power efficiency by reducing capacitance through the elimination of unnecessary circuits.**

Meanwhile, data movement can consume more power than computation. If data needs to be moved in and out of the chip, power consumption can be 3 orders-of-magnitude larger. Therefore, DRAM access which is required by the von Neumann architecture represents a bottleneck in power reduction.

To improve the chip data interface, it is important to shift from a peripheral to an array input/output (I/O) structure. Increase in integration density inside the chip is proportional to the square of the scaling factor. On the other hand, when external I/Os are mainly placed along the perimeter of the chip, increase in I/O density is only linearly proportional to the scaling factor. As a result, the increase in data communication performance cannot keep up with the demand of internal processing. An effective solution is to stack chips on each other so they can be connected across their entire surface. In other words, power efficiency can be greatly improved by moving chip integration from 2 to 3D.

This illustrates how the slowing of Moore's Law offers more and more opportunities for the adoption of disruptive technologies.

3.2   Scaling Scenarios: The Wonders of the Exponential Function

Ideal Scaling Scenario

The fundamental principle driving the evolution of the IC is scaling, which is the miniaturization of semiconductor devices. It improves chip performance while lowering manufacturing costs by increasing integration density.

Integration density has been increasing fourfold every 3 years for DRAM, and twofold every 2 years for processors. Such growth follows the rule of thumb of Moore's Law.

The manufacturing cost of a chip is computed by dividing the cost to manufacture a wafer by the number of good chips on the wafer [22].

Device scaling is achieved through improvement to both lithography and process technology. At the same time, wafer size is increased while manufacturing technology is improved to increase yield, resulting in an increase in the number of good chips per wafer [23].

In the last 50 years, devices have shrunk 20% while chip size increased 14% every two years. Together they resulted in doubling ($= 1.14^2 / 0.8^2$) of the number of devices integrated per chip.

For DRAM, additional efforts including adoption of 3D device structure and circuit improvement led to a total of fourfold increase in integration density every 3 years. Nevertheless, such efforts are approaching their limits. As a result, DRAM scaling is expected to stop in the near future.

Let's now take a closer look at how performance scales. If we **reduce the supply voltage V [V] by the same factor of $1/\alpha$ as is used in shrinking the device dimension x [m]** (a 20% shrink corresponds to $\alpha = 1.25$), the electric field inside the transistor [V/m] remains unchanged [24]. Since transistor operates on electric field effect, this **constant electric field scaling** ensures that the transistor performs the same way before and after scaling.

Under this scaling scenario, both the current I [A] flowing through the transistor and its capacitance C [F] are also reduced by a factor of $1/\alpha$. This is because current is proportional to device dimension. On the other hand, capacitance is determined by area divided by thickness. Since area is reduced by $1/\alpha^2$ while thickness by $1/\alpha$, capacitance is reduced by $1/\alpha$ [25].

When each of voltage, current, and capacitance scales down by the same factor of $1/\alpha$, circuit delay also scales down by $1/\alpha$. This is because delay is computed from the product of capacitance and voltage divided by current [26].

However, power density, defined as power consumption per unit area and computed from the product of voltage and current divided by area, remains unchanged, despite the scaling. Therefore, while one may feel that heat removal will become difficult as integration density rises, power density actually remains the same. **And since the amount of heat generation is close to being proportional to power density, scaling does not result in heat removal problems. It is indeed an ideal scenario.**

Actual Scaling and Its Side Effects

However, in reality, scaling did not follow this ideal scenario.

Microprocessor operating frequency increased 50-fold in 10 years. Scaling contributed a 13-fold increase, while the remaining fourfold increase was achieved through architectural improvement.

This translates into a 1.6-fold increase in operating speed every 2 years, which exceeds the 1.2-fold increase expected from constant electric field scaling.

Device scaling was implemented without lowering voltage up until 1995. In other words, instead of "constant electric field" scaling, "constant voltage" scaling was the reality.

Under this scenario, current I increases by a factor of α. Combined with a C reduced by a factor of $1/\alpha$, circuit delay decreases by a factor of $1/\alpha^2$. As a result, there is an additional increase in circuit speed. However, power density increases rapidly as a function of α^3, resulting in a proportional increase in heat generation [27].

This actual scenario was driven by the desire to generate more revenues through offering even higher performance chips. The increase in power was not too much of a problem since it was very small to begin with.

Consequently, chip power increased 1000-fold in the 15 years between 1980 and 1995. As a result, the amount of heat dissipated in a unit area of the chip reached 30 times that of a hotplate used for cooking.

If the dissipated heat is not completely removed, internal device temperature will rise, resulting in degraded reliability. As scaling hits this power wall, it prevents further increase in integration density.

The power wall was the consequence of overly aggressive scaling.

Eventually, supply voltage started to decrease gradually from 1995.

Furthermore, additional, incremental efforts were made to conserve power, including aggressively turning off power to circuits not in use and lowering supply voltage when high performance was not required.

While these are obvious actions to conserve power in our daily life, it is not easy to identify wastes in a large-scale integrated circuit that consists of more than 100 million transistors.

The theoretical lower limit on supply voltage is 0.036V at room temperature. Below this limit, CMOS circuit gain falls below 1, and it becomes impossible to cascade digital circuits.

However, in actuality, it is difficult to lower voltage beyond 0.45V for multiple reasons, including transistor leakage current in off state, device variations, and noise.

From the 28nm generation onwards, increase in integration density is accompanied by an increase in the number of unusable transistors. In other words, dark silicon (transistors that remain "dark" due to their power being shut off) is increasing rapidly. This means that even though additional functionality can be integrated, it is difficult to actually deliver the added performance.

Therefore, only designers who can improve power efficiency can achieve higher performance. Put differently, there is **no performance improvement without improvement in power efficiency.**

Besides lowering supply voltage, power efficiency can also be improved by reducing capacitance C. Therefore, **technology to stack and integrate chips in 3D is key to the future of IC. In other words, we need to raise integration level from 2 to 3D. Since chip thickness is 3 orders-of-magnitude smaller than chip surface dimensions, the shift to 3D integration can dramatically shorten inter-chip connections, thereby reducing their capacitance.**

The Intuition-Defying Wonders of the Exponential Function

There is a story about an elderly man taking care of the koi fish in a pond. His responsibility was to periodically remove lotus leaves from the pond to ensure that it got enough oxygen for the fish. Since the leaves did not grow that rapidly, he thought it would be fine for him to be away for a week. But when he returned, to his surprise the entire pond was covered in lotus leaves.

This is a story that illustrates the characteristics of exponential growth. (The number of coronavirus infections grows in the same way.)

Our intuition predicts changes by linear extrapolation. We developed this and had it encoded in our DNA in ancient times when we had to protect ourselves from wild animals in the jungle which moved at constant speed. Even in modern days, we frequently predict the future by linear extrapolation of changes that have occurred.

However, the world created by the IC grows at an exponential rate. AI is one such example. It is why AI adoption has been skyrocketing ever since it suddenly burst into existence.

The growth in data volume generated by the IC has also been exponential. The volume of internet traffic is quadrupling every year (Gilder's Law).

By the second half of the 21st Century, the number of transistors that can be integrated on a single chip may rival the total number of neurons in all human beings combined. Furthermore, building a giant brain by wirelessly connecting together all the chips in the world may not be just a dream anymore.

The world is undergoing dramatic changes in less than 100 years since the invention of the IC. (Fig. 3.2).

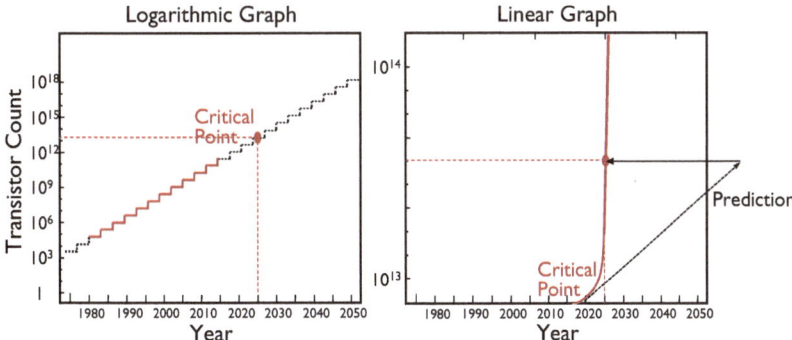

Fig. 3.2 Since technology improves exponentially, it changes at a much faster pace than the linear prediction of human intuition

3.3   Structural Transformation of the Chip: Reducing Leakage

Transistor Structural Transformation

A transistor has three terminals. The first terminal, the source, is the supply port for electric charge; the second terminal, the drain, is the drain port for electric charge; and the third terminal, the gate, is the floodgate that regulates the flow of electric charge. By changing the potential of the gate, the flow of electric charge from the source to the drain can be controlled. In other words, we can make a transistor into a switch.

To make a transistor, the surface of a semiconductor substrate is oxidized to form a thin oxide layer, upon which a metal gate is placed. Next, impurities of opposite polarity to those initially added to the semiconductor substrate are introduced from above. As a result, impurities are implanted into the surface of the semiconductor substrate at the two ends of the gate, forming the source and drain [28].

It is called a MOS transistor because its cross-section is a stacked structure of metal (the gate), oxide, and semiconductor. PMOS refers to a MOS transistor in which the source and drain are formed using a p-type semiconductor which has an abundance of positive electric charge carried by what is known as "holes." On the other hand, NMOS refers to a MOS transistor in which the source and drain are formed using an n-type semiconductor which has an abundance of negative electric charge known as electrons.

NMOS operates as follows. First, electrons gather at the source. When the gate is at the same potential as the source, the p-type semiconductor substrate between the source and drain creates a barrier that prevents electron flow from the source to the drain even if a voltage is applied between the two terminals.

However, when the gate is given a sufficiently higher potential than the source, the surface of the p-type semiconductor substrate immediately below the gate reverses to n-type, creating a channel for electrons to flow from the source to the drain. Since current flows in the opposite direction as electrons, it flows from the drain to the source.

The behavior of PMOS is the exact opposite of that. Holes gather at the source. When the gate is given a potential sufficiently lower than the source, a channel is created that allows the holes to flow out of the source to the drain, resulting in current flow in the same direction.

By connecting the source of a PMOS transistor to the power supply and the source of an NMOS transistor to ground, and connecting their gates together as the input, and their drains together as the output, we can construct a CMOS inverter.

When a low potential (L) is applied to the input of the CMOS inverter, the NMOS turns off while the PMOS turns on, resulting in current flowing from the power supply to the output to produce a high potential (H). Similarly, when H is applied to the input, current flows from the output to ground to produce L at the output.

Since PMOS and NMOS do not turn on simultaneously, there is no steady current flowing from the power supply to ground. In other words, current only flows to switch the output between H and L, resulting in low power consumption. Because PMOS and NMOS operate in this complementary manner, it is called a CMOS.

However, when the transistor gets smaller, leakage current starts to flow between the drain and the source. To understand why, we need to look a little more closely at how the channel is formed.

Let us return to the explanation of NMOS operation. When the gate is given a potential sufficiently higher than the source, why does the surface of the p-type semiconductor substrate flip to n-type, creating a channel?

Let's think of a capacitor (also known as a condenser), which consists of two parallel metal electrodes facing each other. Give one metal plate, let's call it Plate A, a positive charge. The charge will be uniformly distributed across the plate.

Next, let's bring an uncharged metal plate, Plate B, close to Plate A in a parallel configuration. Due to electrostatic induction, the positive charge on Plate A pulls negative charge onto the inner surface of Plate B, while an equal amount of positive charge is induced on its outer surface. As a result, positive charge becomes concentrated on the inner surface of Plate A as well.

Now if we connect Plate B to ground, the positive charge that has been induced on its outer surface will be discharged to ground. However, the negative charge on its inner surface is prevented from moving by the pull from the positive charge on Plate A. Consequently, charge is stored in the electric field within the capacitor.

At this point, replacing Plate B with a p-type semiconductor substrate will give us an NMOS.

When the gate of Plate A is given a potential sufficiently higher than the source, a positive charge is given to the gate and a negative charge is stored on the surface of the p-type semiconductor substrate opposite the gate. When this negative charge, carried by electrons, becomes sufficiently large, the surface of the semiconductor substrate flips to n-type, creating a channel through which electrons can pass.

Thus, the channel is controlled by the field effect from the gate.

However, there is another capacitor lurking in the background, in addition to the gate that affects the channel. It is the drain.

In fact, at the interface between the drain and the semiconductor substrate, electrons from the drain diffuse into the p-type semiconductor substrate while holes from the p-type semiconductor substrate diffuse into the drain. It is similar to the situation where, in a container, if we remove a plate that was previously keeping sugar and salt apart, they will mix together.

The difference is that, due to electrostatic force between electrons and holes, once they mix to a certain extent, they will stop diffusing further.

As a result, a depletion layer is created at the interface between the drain and the semiconductor substrate, which is deprived of electric charge that can move freely. This acts as an insulating layer, creating a capacitor.

When the transistor gets smaller, the distance between the source and the drain becomes shorter, and the depletion layer at the drain gets closer to the source. In other words, the drain also looks like a small gate to the source.

Therefore, even when the gate is closed, if a positive voltage is applied to the drain, the barrier that is trapping electrons at the source will drop slightly, causing leakage.

Even the slightest leakage would become major leakage when 10 billion transistors are integrated.

The cause of the leakage is the weakening of gate control. So, what can be done to improve gate strength?

The first step taken was to change the materials.

The gate oxide layer was changed to a material with a high dielectric constant. This was an effective measure, but its effectiveness was eventually lost as scaling continued.

The next step taken was to change the transistor structure [29].

The structure was changed to one in which the gate is split into two to sandwich the channel from two opposing sides. Since the manufacturing cost would be high to add a new gate under the channel, the channel is turned on its side to stand on the surface of the semiconductor substrate to allow gates to be formed on the front and back sides of it.

That is how we get FinFET (Fig. 3.3 center).

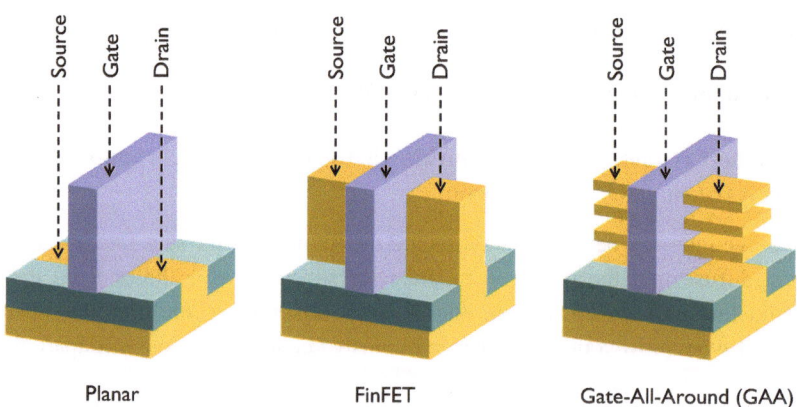

Planar   FinFET   Gate-All-Around (GAA)

Fig. 3.3 Transistor structural transformation

The name FinFET is given to this transistor structure because its shape resembles the fin of a fish. The term "Fin" refers to the vertical channel region, and "FET" stands for field-effect transistor. It has been used since the 16 nm generation.

In the 2 nm generation, a structure with even higher gate strength is necessary. Therefore, a structure in which the gate wraps around the channel is being researched and developed. This is the gate-all-around (GAA) structure (Fig. 3.3, right). A sufficiently large current must flow through the thin channel surrounded by the gate when it is turned on. The required material properties are being studied.

To help you get the picture, let's use a metaphor. You want to stop water from flowing out of a hose. Conventionally, you can do that by pressing down on the hose with your index finger (planar type). However, when it starts to leak, you can pinch the hose from both sides with your thumb and index finger (FinFET) instead. Finally, when that fails too, you can grip the hose with all five of your fingers (GAA).

Structural Transformation of Wiring

As circuit integration increases, the chip consumes more power and generates more heat. Therefore, the maximum allowable power is determined by the maximum permissible temperature rise without causing the chip to fail.

Lowering the power supply voltage is an effective approach to increase integration density without increasing power consumption. By reducing the power supply voltage to half, the power consumption of CMOS circuits can be reduced to one-fourth, allowing for a fourfold increase in integration density. In the 1980s, the supply voltage was 5 V, whereas it is 0.5 V today. Theoretically, the supply voltage can be reduced to 0.036 V at room temperature.

However, a major challenge has arisen—the power supply wiring.

Power is determined by the product of voltage and current. When the integration density increases and the voltage is lowered to prevent power from rising, the result is an increase in current. For example, when the power is 50 W and voltage 5 V, the current is 10 A. However, when the voltage is lowered to 0.5 V, the current will jump to 100 A at the same power.

Microwave ovens and hot plates operate at 10 A. How can we supply a current that is 10 times greater than that to a tiny chip measuring 1 cm^2?

To achieve that, the power supply wiring needs to be made thicker and wider. As a result, power supply wiring on-chip has become thicker and multi-layered, which goes against the trend of scaling. In the 1980s, the number of wiring layers on a semiconductor substrate was 2 or 3, but recently the number has exceeded 15.

The lower layers are used for short-distance wiring, the middle layers for long-distance wiring, and the upper layers for power wiring. The lower layers are thinner and narrower, while the upper layers thicker and wider. Most of the wiring volume is used for power wiring. Various wiring paths are intricately routed throughout the entire chip, similar to the network of blood vessels from capillaries to major arteries in the human body.

The further the transistors are scaled, the thicker and wider the power supply wiring becomes. To solve this dilemma, a structural transformation is about to begin in which power supply wiring is buried in the semiconductor substrate and power is supplied from the backside of the chip.

The Trendiness of Foretelling the End

It has long been predicted that scaling will be reaching its limits.

In fact, such remarks have been made repeatedly since the 1980s. However, the scaling process has continued to defy such predictions to this day.

Some people mockingly refer to it as the "trendiness of foretelling the end." It highlights the irony that the same claim of "reaching the limits in 10 years" has been made many times for over 40 years. Looking at it from a different perspective, however, it can be said that by identifying what would reach its limit in 10 years if we do not do anything, we can conclude that our efforts have helped us overcome such limit.

In fact, replacing the gate oxide layer with a high-dielectric-constant material seemed like an impossible challenge. The ability to fabricate over 10 million transistors with high yields was made possible by the creation of a high-quality gate oxide layer through the oxidation of the silicon substrate surface. Despite the widespread skepticism that it could be achieved, high-dielectric-constant gate insulating layers were eventually realized and put into practical use.

The planar structure of transistors, which has lasted nearly half a century, has undergone a bold structural transformation into FinFET, and eventually GAA. In fact, research on the CFET (complementary field-effect transistor) structure has begun, which stacks GAA PMOS and GAA NMOS on top of each other. From now on, we will have to rewrite semiconductors textbooks every 10 years.

Intel CEO Pat Gelsinger sounded the alarm in 2001.

He observed that, at the time, CPUs had an equivalent power density that exceeded 100 W/cm^2 when calculated at the surface, which is comparable to that of a nuclear reactor. Back in the days of the Pentium CPU, it was at the level of a hot plate. But if things were to continue at the same rate, in ten years, it would be at the same level as the surface of the sun.

Of course, the crisis was averted.

When I was at Toshiba's research laboratory, I learned something from a senior colleague.

"Do not say it is impossible. Even if you think it is impossible now, it may become possible in the future. We should say it is very difficult instead."

I have taken that to heart.

3.4   AI Chips: Learning from the Brain

Computers Born from Mathematics

In ancient times, humans counted by folding their fingers and measured distance by counting their steps. However, humans cannot comprehend large numbers. The invention of computing tools during the time of the four ancient civilizations helped expand human's cognitive capacity.

As mentioned earlier, during the time of ancient Greece, mathematics evolved from being a tool into a way of thinking, and the interior world of mathematics became a subject of research. The invention of symbolic algebra in the fifteenth century during the Renaissance made it possible to inquire about the nth dimensional space which is hard to represent in the physical world. In such manner, mathematics achieved widespread application without being bound by physical constraints.

Eventually in the seventeenth century, calculus was invented which enabled the study of the world of infinite. After close examination of the concepts of limits and continuity, abstract symbolism was born which exceeded what subjective intuition could achieve. Then as we entered the twentieth century, efforts were started to make mathematics about how to perform mathematics.

As a result, **mathematics moved out of our body into our brain, completely departed from ambiguity represented by the likes of physical intuition and subjective feelings, and eventually flowed out of our brain to give birth to the computer.**

Early electronic computers were plagued by frequent breakdowns of the vacuum tube. Vacuum tubes are devices that control the flow of electrons emitted into a vacuum from a heated electrode. Similar to household incandescent bulbs, the electrodes in vacuum tubes would also grow thinner gradually over time and eventually break.

The problem was solved in 1948 by the invention of the transistor which manipulates electrons in a solid medium instead of empty space. This resulted in a jump in the reliability of the device.

Furthermore, as explained in Chapter 2, wired logic used in early computers where the wiring determined the computer's functionality suffered from two challenges. **The first was the challenge of scale where the size of the largest programs that could be processed were limited by the scale of the available hardware. The second was the challenge of wiring where the number of interconnections exploded as the scale of the system grew.**

To overcome the first challenge, von Neumann invented the **stored-program architecture (von Neumann architecture)**, in which a single arithmetic unit executes a different instruction each cycle, instead of having multiple physically connected arithmetic units. This innovation resolved the challenge of scale.

Meanwhile, Jack Kilby invented the **integrated circuit (IC)** in 1958. The challenge of wiring was overcome by using photolithography to simultaneously integrate many devices and wires onto a single chip. Eventually, silicon was found to be the best material for fabricating ICs [30].

In this manner, integration and parallelization of simplified and miniaturized computing resources onto silicon chips resulted in a dramatic improvement of computer performance. In turn, computers with higher performance enabled the design of even larger scale integrated circuits.

As a result, **the combination of von Neumann architecture, IC, and silicon enabled the co-evolution of the computer and chip to deliver exponential advance.**

Energy is consumed while doing work. The amount of work that an IC can do, in other words, the IC's performance, is limited by the power supplied and the heat that can be removed. Therefore, chip performance can be improved by improving energy or power efficiency, where power is the rate of energy flow.

Chip power efficiency has improved by three orders of magnitude in the last 20 years, rising to 1/100 times that of the brain. At the same time, the level of chip integration has reached 1/100 times that of neurons in the brain. If these trends continue, the chip can catch up with the brain in 10 years.

However, because the von Neumann architecture requires a large amount of data and commands to flow back and forth between the processor and memory, that interface has become a bottleneck (the von Neumann bottleneck). Furthermore, as device dimensions dropped below 100 nm at the beginning of the twenty-first century, quantum effect began to surface, and silicon chip leakage current cannot be suppressed anymore. The growth of the computer and chip born half of a century ago is approaching its limits.

But before reaching its limits, the computer acquired the ability to learn on its own. It is machine learning. And AI chips which mimic neural networks of the brain were born.

AI Chips Learning from the Brain

Although the underlying technologies required for the design of neural networks were developed during the twentieth century, the vast solution space made it difficult to train deep neural networks with more than 4 layers.

However, as we entered the twenty-first century, with the successful creation of deep autoencoders and the achievement of sufficiently large computer performance required for training, deep learning became capable of delivering overwhelmingly larger processing performance than conventional processing, resulting in its rapid adoption [31].

Research on both network structure and architecture has also advanced. CNN (convolutional neural network) for image recognition where feature extraction is localized was successfully developed. Furthermore, research was conducted into networks for recognition processing of time series data such as voice and natural language using RNN (recurrent neural network) and LSTM (long short term memory). More recently, the transformer architecture which uses a self-attention mechanism to focus attention on the important parts instead of using the recurrent construction of an RNN has gained prominence.

All these research efforts take their hint from the human brain. Especially important among them is the **pruning of neural networks**.

Although we are born with only about 50 trillion synapses in our brain, a year after birth the number grows to 1000 trillion. However, thereafter, as we learn, the number of synapses decreases. While synapses with signal passing through get strengthened and remain, unused synapses which receive no signal are pruned and disappear. By around age 10, the number of synapses is reduced to half, and remains roughly the same afterward.

In other words, **while the neural network that forms in our brain during our early childhood is close to a fully coupled network, as we learn, unnecessary wiring is removed. This is how we develop a lean and efficient functional neural network in our brain.**

Children have a large brain to facilitate learning. But an adult's brain is pruned in order to perform inference efficiently. Born small, raised to grow large, and made to learn in society is the survival strategy of mammals equipped with a well-developed brain.

Brain and Silicon Brain

Let's summarize our discussion so far regarding both the human and silicon brain. **Born out of mathematics, the von Neumann computer is able to perform robust information processing based on pre-programmed state transitions. It works like the thalamus, amygdala, and cerebellum of the human brain whose functions are determined through inheritance.**

On the other hand, wired logic-based neural networks, taking hints from the human brain, continue to learn as an open system while being pruned, and perform open-loop, flexible information processing with high energy efficiency. They work like the cerebral cortex which learns and develops in a societal setting.

Silicon brain can thus be described by referencing to the human brain (Fig. 3.4). Does that mean silicon brain will assume the same structure as the human brain?

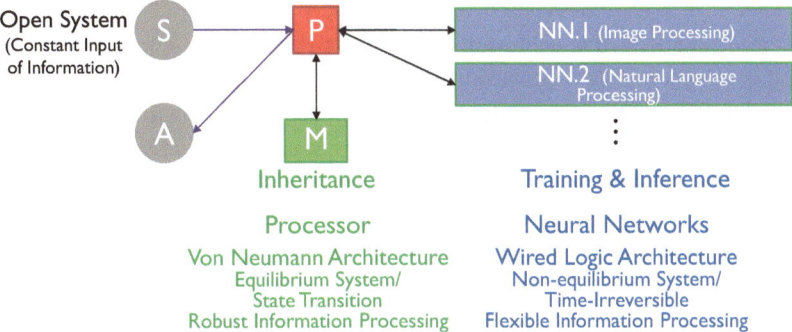

Fig. 3.4 Silicon brain. Processor serves the role of the thalamus, amygdala, and cerebellum, while neural networks serve that of the cerebral cortex. (S: Sensor, A: Actuator, P: Processor, M: Memory, NN: Neural Network)

"That is an amazing dynamic range!" exclaimed Mr. Kazuyuki Aihara in 1981, who at the time was a senior colleague of mine in my research lab (and is now a professor at the University of Tokyo). He was referring to the large variation in resistivity of the nerve axon he found by simulating and analyzing it using the Hodgkin-Huxley model, which is a set of nonlinear differential equations that describes the initiation and propagation of action potentials in nerve axons.

It will not be easy to artificially create something with matching characteristics. The human and silicon brains may end up having different structures based on different principles, just like how birds and airplanes have become.

Neural networks have a wired-logic architecture where functionality is determined by the wiring. As a result, I have great expectations for FPGA (field-programmable gate array) because of its programmable wiring.

3.5   [Column] LSTC's Strategy

The Leading-edge Semiconductor Technology Center (LSTC) was established as a Collaborative Innovation Partnership (CIP) [32] on December 21, 2022.

Its mission is to establish an open R&D platform and formulate technological strategies to design, develop, manufacture, and research next-generation semiconductors with node size smaller than 2 nm, with a focus on achieving short TAT (turnaround time) for mass production. Short TAT means reducing the time taken from the start to the completion of development or production.

The specific development tasks include tools and methodologies for circuit design and verification that enable short TAT, innovative semiconductor device technologies that surpass the performance of conventional semiconductors, manufacturing and measurement techniques that contribute to short TAT and the development of semiconductors with node size smaller than 2 nm, materials that enhance the performance of semiconductors, 3D packaging technologies that achieve both hardware performance improvement and short TAT, and exploration of new devices that have the potential to create new industries.

To accelerate its development efforts, LSTC will actively collaborate with the National Science and Technology Council (NSTC) in the U.S. and imec in Europe.

At the same time, LSTC will work to develop semiconductor professionals for sustained prosperity of the semiconductor industry.

LSTC members include the National Institute of Advanced Industrial Science and Technology (AIST), RIKEN, National Institute for Materials Science (NIMS), Tohoku University, the University of Tsukuba, the University of Tokyo, Tokyo Institute of Technology, Osaka University, and the High Energy Accelerator Research Organization, which are responsible for research and education in science and technology related to semiconductors. And Rapidus, responsible for mass production, is also a participating member.

There are two prior examples of a collaborative innovation partnership that can be used as a reference in considering LSTC's strategy.

The first example is the Very Large-Scale Integration (VLSI) Technology Research Association. It operated from 1976 to 1980 and played a significant role in the rise of Japan's semiconductor industry in the 1980s.

Fujitsu, Hitachi, Mitsubishi Electric, Tokyo Shibaura Electric (now Toshiba), NEC, Nichiden Toshiba Information Systems, and Computer Research Laboratory joined forces to research and develop solutions to two technological challenges common to all companies: development of manufacturing equipment for VLSIs and manufacturing technology for large-diameter, high-quality wafers.

They developed the reduction projection exposure system (also called a stepper) that dominated the global market and contributed to a jump in the domestic production ratio of semiconductor manufacturing equipment from 20 to 70% [33].

The method by which engineers from competing companies worked together on a common technical challenge became a global model.

The second example is the Research Association for Advanced Systems (RaaS).

Amid the decline of Japan's semiconductor industry which started around the year 2000, the preservation of Japan's technology and technical talent and the formulation of strategies are important steps towards its revival.

The strategic goal of RaaS is to improve both energy and development efficiency. This strategy, born in response to the trends of the times, is linked to the strategies of Rapidus and LSTC.

RaaS was launched in 2020 with strong support from the private sector. Toppan Printing, Panasonic, Hitachi, and MIRISE Technologies are the founding members of the organization. It is the result of business leaders, who recognize the power of hardware, making a bold decision against the tides amidst a series of withdrawals from the semiconductor business.

In addition, the Japanese subsidiaries of various foreign corporations including EDA vendors and foundries, worried about the decline of Japan's semiconductor industry, reached out to their headquarters to win their complete support for RaaS.

Thanks to the collective effort of its members, RaaS was able to create a cutting-edge design environment for the 7 nm node, which was state-of-the-art at the time.

In addition, Advantest and RIKEN joined RaaS in April 2023 as the newest members.

Given how semiconductor technology can contribute to the advancement of science, as previously noted, RaaS is pushing forward the democratization of semiconductors.

Meanwhile, in order to improve energy efficiency, a NEDO (New Energy and Industrial Technology Development Organization) project that focuses on the research and development of 3D integration technology utilizing Japan's pool of superior technologies was initiated in 2021. SCREEN Holdings, Panasonic Connect, Daikin Industries, and Fujifilm have joined the project to develop technologies for establishing a strategic advantage.

There is much for LSTC to learn from these two collaborative innovation partnerships.

First, we need to come together in an international collaborative effort with Japan and the U.S. at its core to promote research and development focused on improving both energy and development efficiency, which are common challenges for all humankind.

Second, creating a rich industrial ecosystem and promoting co-existence and co-evolution of those who participate will lead to sustainable and flexible industrial development.

To that end, we need to mobilize everyone to join the academic pursuit of diverse disciplines from design to device, manufacturing, equipment, and materials, in order to achieve total optimization instead of piecemeal, partial optimizations [34].

Finally, technology is people. In other words, human resource development is essential for lasting development. Therefore, it is important to create an attractive open platform to bring the world's brains together.

The sun will rise again.

The mission of LSTC is to refine Japan's technology and develop its human resources and formulate strategies in preparation for the dawn.

It is interesting, even if coincidental, that both the VLSI Technology Research Association and LSTC were born at the dawn of the era of specialized semiconductors.

A Fertile Ground for Innovation: More Than Moore

4

4.1 From 2 to 3D: The Next Half-Century of Integrated Circuits

Connectivity Problems in Large-Scale Systems

The integrated circuit (IC) was invented against the backdrop of a connection problem—the challenge of wiring—in large-scale systems.

ENIAC, one of the earliest computers developed in 1946, contained about 5 million handmade connections. As systems scaled up, the number of connections exploded.

This problem was known as the **tyranny of numbers**. After trying to solve the problem using various approaches, the definitive solution that emerged was the IC.

Since then, the IC has undergone exponential improvement following Moore's Law, with computer performance increasing dramatically in lockstep.

However, due to large data movement between memory and processor, inter-chip communication has become the main reason for the erosion of energy efficiency resulting in the **von Neumann bottleneck**.

With the combined effect of data explosion, the industry has fallen into a situation where there is **no computing performance improvement without improvement in energy efficiency**. The problem has persisted to this day.

CMOS circuit energy consumption is proportional to its loading capacitance. The loading capacitance of computational circuits can be reduced through device scaling.

However, since data movement involves charging and discharging the capacitance of the entire communication path, even if devices are scaled down, energy consumption cannot be reduced without shortening the communication distance.

Data movement can consume a lot more energy than computation.

For instance, compared to 64-bit data processing, moving data to the edge of the chip and then out of the chip to DRAM require 50 and 200 times more energy respectively.

© The Author(s), under exclusive license to Springer Nature Switzerland AG 2025   49
T. Kuroda, *The Super-Evolution of Semiconductors*, Synthesis Lectures on Engineering, Science, and Technology, https://doi.org/10.1007/978-3-031-60518-5_4

Another reason that has contributed to the large energy consumption of inter-chip communication is the aggressive increase in transmission data rates. This was due to the placement of communication channels along the chip perimeter only, making it difficult to increase the number of channels.

IC computing performance has been increasing 70% annually, as a result of a 15% transistor speedup combined with a 49% increase in functional integration density.

To take full advantage of this chip functional improvement, the speed of data movement in and out of the chip must increase accordingly.

Using the rule of thumb known as Lent's Rule that infers the increase in the number of communication channels (chip I/Os) required to keep up with the increase in the scale of computing logic, the inter-chip communication speed needs to increase at a 44% annual rate.

However, device scaling can yield only a 28% annual increase in inter-chip communication speed. This is because transistors speed up by 15% while the number of I/Os increases only 11% since I/Os are placed only along the chip perimeter.

Even if I/Os are placed across the entire surface of the chip, signal trace congestion will make it difficult to route all the signals out of the chip on the printed circuit board without using a lot of board layers.

As a result, innovative circuit technologies have been used to raise inter-chip data rates to close the gap. Unfortunately, as is generally true, this aggressive push of transistor performance to its limits requires consumption of a lot of energy.

The energy consumed by inter-chip communication started to rise with the 130 nm generation (around the year 2000). We are now approaching the limits of how much more we can increase communication speed.

We can thus conclude that in order to increase the energy efficiency of computing, we need to shorten the distance between memory and processor, as well as increasing the number of connections to avoid pushing up data rates too aggressively.

In other words, we should stack chips in 3D to minimize inter-chip distances and to allow the entire chip surface to be used for interconnections to enable a moderate data rate. This is the reason why we are transitioning from 2 to 3D chip integration.

Without being able to rely solely on on-chip integration for performance increase, we have been evolving from 2 to 3D chip integration, which calls for a breakthrough solution to the connection problem. [35]

Through Silicon Via and ThruChip Interface

Against this backdrop, the research and development of TSV (through-silicon via) for vertically connecting stacked chips was started in 1990. Compared to conventional processing which requires penetration of only a few microns from the chip surface, TSV processing is challenging because it requires penetration of a few tens of microns.

Furthermore, solder connections are difficult to shrink, and there are stress and reliability problems arising from thermal coefficient of expansion mismatch between materials.

To this day TSV incurs high costs and lowers reliability. More than a quarter century later there is still no solution on the horizon.

In recent years, there has been significant progress in wafer bonding technology that directly connects copper electrodes together without using solder. This is called Cu–Cu direct bonding. Alternatively, it is also called hybrid bonding because of the presence of both copper electrodes and a silicon oxide layer on the bonding surface.

The alternative is to use **magnetic coupling for inter-chip communication as in ThruChip Interface (TCI)**. With TCI, communication is achieved by using the digital signal to be transmitted to control the direction of current flow in coils formed on the interconnect layers of the transmitting chip to alter the direction of the associated magnetic field, which in turn changes the polarity of the signal induced in matching coils on the receiving chip. The induced signal is then detected and used to regenerate the original digital signal in the receiver [36]. In other words, communication is achieved through magnetic coupling between coil pairs.

Since all materials used in semiconductor chips have a relative permeability of 1, magnetic field can readily penetrate through chips. Furthermore, since CMOS circuits operate on electric field effect, there is no need to worry about interference.

However, **TCI's biggest merit is that its electrical connections are formed using wafer process and standard CMOS circuits, as opposed to the mechanical connections of TSV which are formed in packaging and assembly process**.

Since TCI is implemented using digital circuits without modification to the chip manufacturing process, it can be realized at low cost by anyone. While TSV may increase DRAM cost by 50%, with TCI, the cost adder can be kept below 10%.

Moreover, as the chip thickness is reduced, the cost performance of TCI can be dramatically improved.

For instance, if the chip process is scaled by 1/2, and in addition the chip thickness is reduced by 1/2, TCI's data rate can be increased eightfold while its energy consumption reduced to 1/8.

However, TCI cannot be used to deliver power. The current solution is to use TSV for power delivery and TCI for signal connections. You may wonder if it is not easier to just use TSV for signal connections as well. The reason is that TSV failures are mainly open failures. As a result, TSV failures can be solved with redundant connections. While it is difficult to add redundancy to signal connections, redundancy is built into the power delivery network since it consists of highly parallel connections.

Research and development of a new power delivery technology using highly doped regions of impurities known as highly-doped silicon via (HDSV) has also commenced as a replacement for TSV.

Fig. 4.1 In-package 3D integration of memory and processor improves energy efficiency

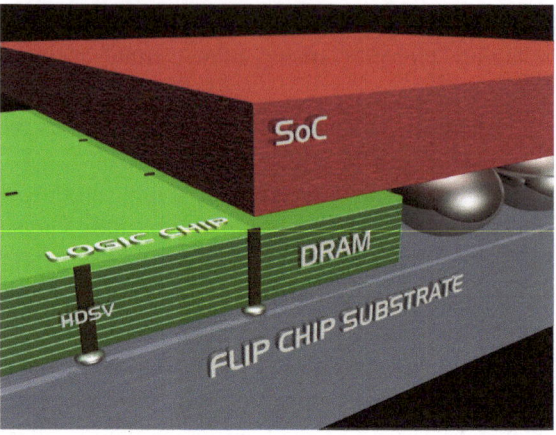

The transition from 2 to 3D integration results in higher chip energy efficiency. However, as chips are stacked on top of each other, the power density of the stack rises. Consequently, it is necessary to further improve power efficiency to avoid generating more heat in the stack than can be removed. (Fig. 4.1).

An Era of Disruptive Technologies
In startup jargon, there is a valley of death stretched between the research and adoption stages. It is difficult for a disruptive technology to cross the valley to achieve adoption.

For a connection technology, both sides of the connection must agree to adopt the technology.

When introduced to a processor company, even if TCI generates interest, the next question that arises is when memory companies will adopt it.

When the fact that the processor company shows a lot of interest in TCI is communicated to the memory companies, they respond with a stern face that until all their major customers ask for it, it is difficult to adopt the technology since it requires major investment. Given that memory is a commodity business, the memory companies generally tend to be conservative.

It is hard to resolve this chicken-and-egg problem.

However, the challenge of no computing performance improvement without improvement in energy efficiency **has necessitated the transition into a new era of 3D integration, which makes the time ripe for the adoption of disruptive technologies (which we would like to call revolutionary).**

Nevertheless, it remains difficult to motivate the memory companies. Therefore, a better approach is to start by stacking SRAM chips in 3D to offer memory capacity to processors which can rival that of DRAM. Since SRAM can be developed by the processor company, the adoption decision can be made by the processor company alone. Meanwhile, DRAM scaling appears to be nearing its end.

4.2   Semiconductor Cubes: From Horizontal to Vertical

Pancake Stacking and Sliced Bread Stacking

Three-dimensional integration began with memory.

First, HBM (high bandwidth memory) with two stacked DRAM chips was introduced to the market, followed by an increase in the number of stacked chips to 4, 8, and then 12.

Next, 3D stacking of memory and logic chips was also started: a processor with two SRAM chips stacked on top was announced by Advanced Micro Devices (AMD). The processor's power performance can be improved by 30% by increasing the cache memory capacity in this manner. This performance improvement is equivalent to one generation of process scaling. It truly is More than Moore.

In both cases, the chips are stacked flat. No one arranges them vertically. The chips are 1 cm on the side and about 0.1 mm thick. No one would think of placing such thin chips vertically, at least not when only about 10 of them are integrated together.

But what happens when we integrate a hundred of them? The total thickness becomes 1 cm. In other words, it becomes a cube.

Consider a memory cube with 100 memory chips stacked on top of a logic chip. There are two possible integration schemes for the memory chips in the cube. One is horizontal and the other vertical. Let's call the former the pancake and the latter the sliced bread (as in a pre-cut loaf of bread) scheme (Fig. 4.2).

In fact, the sliced bread scheme has more advantages than the pancake scheme.

First, heat can escape more easily.

The silicon substrate of the chip conducts heat 150 times better than the silicon oxide layers used for wiring insulation. In the sliced bread scheme, the silicon substrate allows heat to escape from the bottom to the top. In the pancake scheme, the silicon oxide layers impede heat flow like multiple layers of blankets.

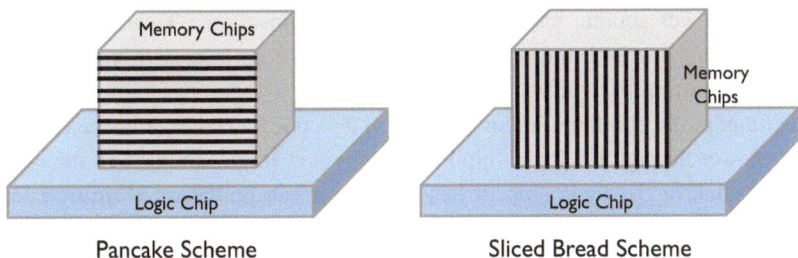

Fig. 4.2 The sliced bread scheme with chips vertically placed in a cube has better heat dissipation, power delivery, and communication performance

The biggest challenge in 3D integration is heat removal.

When many chips are stacked on top of each other, the amount of heat generated increases in proportion to the number of chips. The challenge is how to move this heat to the heat sink on top of the package. In this respect, the sliced bread scheme is much more advantageous than the pancake scheme.

For example, in the case where the maximum temperature inside the cube reaches 200 °C in the pancake scheme, the temperature can be suppressed to 100 °C in the sliced bread scheme.

Second, communication is easier.

In the pancake scheme, memory chips are on top of each other, so the memory chips placed lower down have more penetrating wires connected to the logic chip.

If possible, we want to stack the same memory chips. Therefore, we connect them with a bus, i.e., we skewer the chips and connect all of them to the same wires. But then even when the logic chip needs to communicate with only one memory chip, it must also transmit to all other memory chips, resulting in unnecessary increase in delay and power consumption.

However, in the sliced bread scheme, all memory chips are in physical contact with the logic chip, so direct individual connections are possible even when stacking the same memory chips, minimizing delay and power consumption.

However, special communication techniques are needed to communicate with the logic chip at the edge of the memory chips. Magnetic field coupling makes this possible. The technology communicates by coupling the magnetic field formed around signal wiring. The advantage of this technology is that communication is possible even when the two chips are slightly misaligned.

So, what about power delivery?

If power is delivered from the logic chip to the cube, the pancake scheme seems to be simpler than the sliced bread scheme.

However, as is with communication, when power is delivered from the bottom to the top in the pancake configuration, more current flows through the lower chips which are closer to the power source, requiring more power connections. Therefore, if the same memory chips are stacked, the upper chips would have an excessive number of power connections, resulting in waste.

In addition, as the number of memory chips in the cube increases, it becomes inefficient to deliver power through the logic chip because the power consumption of the cube is no smaller than that of the logic chip. Instead, if power can be delivered directly from the package to the side walls of each memory chip, the size of the logic chip can be reduced. **This line of thinking requires a new technology to deliver power directly from the package to each memory chip**. Various ideas are currently being considered.

By the way, how hard would it be to stack more than 100 chips?

It is relatively easy if you do the following.

First, the front sides of two chips are bonded together and then one of the back sides is thinned to make a module. Next, the thinned sides of two modules are bonded together and then one of the back sides is again thinned. This produces a module with four stacked chips. By repeating this process of doubling the number of chips 7 times, we achieve a stack of 128 chips. Repeating it 10 times, more than 1000 chips can be stacked in the same way.

From Memory Cube to System Cube

Memory includes SRAM, DRAM, and NAND flash. SRAM is the fastest in terms of operating speed, while NAND is the slowest. On the other hand, NAND has the largest storage capacity while SRAM has the smallest. No one memory type is superior in all respects.

Therefore, a memory hierarchy is used. That is, information likely to be needed immediately is placed in SRAM or DRAM. On the other hand, information that will not be used for a while is stored in NAND or DRAM.

If SRAM could be made as large as DRAM, computing performance would increase dramatically, and a revolution would occur.

AMD showed that potential: stacked implementation of two SRAMs on top of the processor resulted in a performance improvement comparable to one generation of process scaling.

So, what happens when 128 SRAMs are stacked?

If the dimensions of the SRAM chip are 8.4 mm × 3.0 mm and 0.1 mm thick, the size of the cube becomes 8.4 mm × 3.0 mm × 12.8 mm.

Using the most advanced process after N2, the storage capacity becomes 24 GB, which is the same as HBM3 with 12 stacked DRAMs. Power consumption is also the same as HBM3.

On the other hand, the data transfer bandwidth is 14.4 TB/s, which is 17 times that of HBM3. Latency is fewer than 10 cycles, which is <1/5 that of HBM3.

If such an SRAM cube can be created, it can replace HBM3 and significantly improve computing performance.

If the thickness of the SRAM chip is reduced to 0.025 mm, the memory capacity can be expanded by a factor of four.

Like SRAM cubes, DRAM and NAND cubes can also be made. By gluing chips together, any type of chips can be made into a cube.

Furthermore, memory cubes can be freely created by combining SRAM, DRAM, and NAND in any desired ratio. The optimal combination can be made for each application.

Logic chips can also be intermingled in the cube to make a system cube.

For example, first, a logic chip that accelerates AI processing is glued together face-to-face with an SRAM chip. In addition, as many DRAM and NAND chips as necessary are added to form a cube. Finally, a logic chip manufactured using a relatively inexpensive process that integrates peripheral, control, and internal networking circuits is mounted under the cube to create a system cube.

Looking at a bookshelf, one sees a mix of books stacked vertically and horizontally. In the future, with a similar arrangement of chips, the inside of a package will probably look like a bookshelf.

The Best Thing Since Sliced Bread

Today, it is commonplace for a loaf of bread to be sold in sliced form. But when bread in sliced form was first introduced 90 years ago, it was revolutionary and became very popular.

Hence, the English idiom "The best thing since sliced bread" was born. The phrase means "something groundbreaking, something wonderful."

For example, one may say "This new smartphone is the best thing since sliced bread!".

With 3D-integrated chips made in sliced bread form by turning them from horizontal to vertical, it is fitting to say,

This new 3D chip stack is the best thing since sliced bread!

4.3   Connecting the Brain to the Internet: The Internet of Brains

Mysterious Sight Seen in Cambridge

The year was 2019. Spring came late to Cambridge, Massachusetts. Even though it was May already people were still wearing thick coats.

As dusk fell, the beauty of the Harvard campus became even more prominent. The crowd of students crossing the freshly green lawn was getting thin, while orange incandescent light from dormitories started to permeate the air. As darkness set in on the campus, I felt the weight of its long history like a curtain coming down on me.

This is where what humankind has discovered is passed on to the next generation, and then new knowledge is born. Stimulated by the atmosphere, I felt an urge to study there. With deteriorating eyesight, I need to make an effort even to just read a book. Yet, I felt that it would be fulfilling if I could keep repeating and continue to learn in school my entire life.

But if that was allowed, school campuses would be overflowing with senior citizens. "Only if I had been able to visit this place while being a little younger ..." was the sentiment that overwhelmed me.

The next day was for meetings on AI chip research. I needed to go to MIT in the morning and Harvard in the afternoon. From The Charles Hotel in Harvard Square, it takes only 15 min by the Red Line subway to go from the nearby Harvard Station to Kendall Station, where the MIT Media Lab is located. But instead, I decided to take a walk.

Charles River was not visible from where I was. Still, I walked leisurely while the boat racing scene from the movie *The Social Network* ran through my head.

Unfortunately, I was not able to find anything interesting along the way as I had hoped. After walking for almost an hour and just when I was getting tired, I finally came to the intersection between Main Street and Vassar Street, which was close to my destination.

At that moment, a magical illustration suddenly jumped out at me.

It was being projected on a large screen set up in the entrance hall of a contemporary building that was radiating blue light due to reflections off its glass windows. "McGovern Institute for Brain Research" was written on the building.

I entered the building while looking up at a monument in the shape of a twisted, giant tree and sank myself into a sofa. There was a security gate in the back. Young researchers were hurrying in and out while holding a smartphone or a cup of coffee in one hand.

These were talented people from around the world. You could see the spirit and self confidence in their eyes, which is common to people on the forefront of the world's most advanced research.

What was being shown on a 100″ display was a slideshow that introduced the Institute's research.

This is it!

The magical image that caught my attention was being projected there.

It resembled both an astronomical photograph and an abstract painting. It was a microcosm composed of numerous curled threads that radiated rainbow-like color in the dark. It looked like sperms in formation charging into an egg.

From its title of "New Image of The Brain," I realized that it was an illustration of the brain's neural network. It was a blueprint of the brain which you could freely alter by changing your viewpoint in 3D.

"Professor Boyden's lab has developed a technology for imaging the interior of a brain cell including its protein and RNA," the slide read.

Then came the next slide which depicted a scientist holding a preparation (a glass slide for microscope specimen) in his hand that was radiating a phosphorescent light. It was entitled "Expansion Microscopy."

The Expansion Microscopy Method and the Opposite Approach

Expansion microscopy?

Is that something that can enlarge cells and tissue? Are we talking about Alice in Wonderland syndrome?

Googling "expansion microscopy," I found an article entitled "Blown-up brains reveal nanoscale details" published in the January 2015 issue of Nature (vol. 517, issue 7534).

The lead sentence read "Material used in diaper absorbant can make brain tissue bigger and enable ordinary microscopes to resolve features down to 60 nm."

Details of the technique were explained in the text. First, protein of brain tissue is tagged with fluorescent molecules. Next, acrylate is infused into the brain tissue to be combined with the fluorescent molecule tags. After polymerization, the acrylate polymer forms a mesh within the brain tissue.

After decomposing the protein in the brain tissue, water is added to the remaining acrylate polymer. As the polymer absorbs water, it swells like a diaper, which causes the separation between fluorescent tags attached to the mesh to grow in all directions. As a result, the fluorescent tags which were previously too close to each other to be distinguishable under an optical microscope are now individually visible.

In other words, it is like copying the positions of brain tissue's protein to a paper diaper and then swelling the diaper with water to enable them to be observed using an optical microscope. The rainbowlike illustration in front of me was a vivid 3D computer rendering of the image. It was an amazing visualization.

In the slide show, Prof. Ed. Boyden asked, "If you want to see the brain better, what would you do? You can shrink the scientist, or you can enlarge the brain tissue."

Of course, Prof. Boyden has chosen the latter.

But me, I would shrink the scientist!

This is when my imagination started to run wild. Let's create a small microscope by integrating 1 million sensors onto a 100 μm^2 chip. Each sensor is thus 100 nm^2 in size.

If we can insert this chip into the brain tissue to observe it from a close distance, shouldn't we be able to distinguish features that are 60 nm apart? By using many such chips to capture images and collecting the data wirelessly, shouldn't we be able to recreate the whole picture? Letting my imagination continue to run free, I forgot about both time and my fatigue.

At the time I was working on an ACCEL project sponsored by the Japan Science and Technology Agency (JST). The original focus of the research was to improve computer power efficiency. But with the AI boom, I had started to contemplate creating mobile AI "eBrains."

If we can embed small chips into the brain, we can connect the brain to the internet. **We can then realize an Internet of Brains (IoB) to succeed Internet of Things (IoT).**

Further down the road, maybe we can move from the brain to the cell and create an Internet of Cells (IoC).

Maybe not. Before that, **we should create a human intranet** where sensors and actuators worn by a person are connected to their brain. **A computer that is integrated into the human brain and body should be able to expand our ability including our senses and immunity and support the social life of senior citizens**.

This has become my dream.

When the Brain is Connected to the Internet
Brains and computers are tightly connected.

The brain created society and gave birth to the mind. Humans acquired languages and logic in order to recognize and communicate their intentions.

Furthermore, as a tool for expanding our cognitive capacity, we developed mathematics. Eventually, high level abstraction of mathematics allowed us to compute in our head instead of using our body (hands and fingers), which led to the birth of the computer.

Like how Dr. Ichiro Tsuda expressed the universality of consciousness by saying "the mind is all mathematics," or as Mr. Masao Morita described in his book *The Body That Created Mathematics*, the computer and AI were born as a result of abstraction.

If we can create an eBrain which includes both a left and right halves just like the human brain, wouldn't the left brain be able to abstract images and sound into words in its association area after they have been detected by the right brain?

If the brain is connected to the internet, and its ability to innovate grows as the number of people connected increases, will ideas reproduce so fast as to completely overwhelm the Earth, like the way Matt Ridley described in his book *The Rational Optimist*? [37]

And then will the aggregation of agents give birth to The Society of Mind as Marvin Minsky advocated? Consciousness and art were born after words. Will computers evolve in the same way as human beings?

(Or will Dr. Takeshi Yoro, an anatomist famous for his book *The Wall of Stupidity*, laugh it away by calling it stupid?).

4.4   To Be Synchronous or Asynchronous: Rhythm of the Chip

Synchronous Design of Chips

Half a century ago, there was an intense debate about the pros and cons of synchronous vs asynchronous circuit design. The former uses a clock to synchronize circuit timing while the latter does not.

The following experiment was conducted at Caltech. The high achievers among a group of students were asked to complete an asynchronous design **while the rest a synchronous design of the same chip. The result, while many of the synchronous designs**

functioned correctly, the asynchronous designs did not. What do you think happened with asynchronous design?

Let's consider a state transition as an example. When a state represented by 2 bits {S1, S2} transitions from {0, 1} to {1, 0}, it may instantaneously be in a transient state of {0, 0} or {1, 1}. That is because S1 and S2 are generated by different circuits. Even if the two circuits are designed to be the same, it is difficult to synchronize their timing because of manufacturing tolerances creating differences in the individual circuit elements.

That instantaneous wavering, known as a dynamic hazard, will not cause any problem in computation because the final answer is still correct. However, in control it can result in errors. That is because if data happens to arrive at the instant of wavering, an incorrect control action can result.

But if we make data that arrives early wait for data that arrives late, and then release them all together at the transition of a clock signal, like how a traffic signal makes all cars start at the same time when it turns green, we synchronize their timing to the clock cycles.

We can hold data by using a loop consisting of two inverters. For example, if we input L into the first inverter, it outputs H, which is converted to L by the second inverter to feed back into the first inverter.

If we insert a switch into the loop that is controlled by a clock, such that the loop is closed to hold data when clock is L, and opened to allow data to pass through when clock is H, we create what is known as a latch to latch onto data.

If we connect two latches together and feed to the first latch a clock with opposite phase, we get a flip-flop. **When clock is L, the first latch lets new data in while the second latch holds current data. When clock switches H, the first latch holds the new data while the second latch allows it to be read out. Therefore, at the instant clock transitions to H, different flip-flops are made to output data in synchronization (Fig. 4.3).**

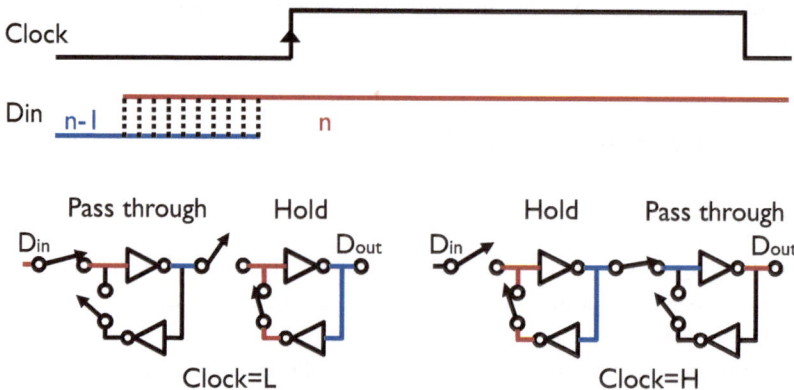

Fig. 4.3 Flip-flop circuit: data is read out on a rising clock edge after being held for a cycle

By the way, when clock switches from H to L, before the first latch lets new data in, the second latch latches onto and hence holds current data. As a result, the flip-flop holds its output at the current value while waiting for new data.

With the use of flip-flops, timing can be verified on a per cycle basis, reducing the cost of verification. Flip-flops are used in general chip design.

On the other hand, when latches are used, data can pass through as long as clock is H, so it is possible to recover from a delay in an earlier cycle. However, to verify timing, one must check not only the current but also earlier cycles, making verification more costly. Latches are used in processor design.

Rethinking Asynchronous Design

Careful timing design is essential to achieve high chip performance. It is necessary to guarantee by design the timing margin required to achieve the target manufacturing yield, by considering the effect of external factors including manufacturing tolerances and variations in power supply and temperature on signal transmission delay of logic circuits, as well as jitters in clock generation and skew in clock distribution.

The required margin grows as process is scaled and supply voltage lowered. Furthermore, as clock speed increases, the cost of timing design, which is known as timing closure, goes up as well.

On the other hand, in synchronous design, since the clock period is determined by the slowest circuits, most of the other circuits do not affect performance. However, clock distribution and flip-flops alone consume 25–50% of total power.

As the cost of and waste in synchronous design became dominant, around the time when clock frequency crossed over 1 GHz, research was started to reconsider asynchronous design. The paper by Ivan Sutherland entitled "Computers without Clocks—Asynchronous chips improve computer performance by letting each circuit run as fast as it can" was published in 2002. Part of Sun Microsystems' UltraSPARC IIIi adopted asynchronous circuits.

Sutherland is a genius who is considered the father of computer graphics. He is also skilled in chip design. He advocated the use of logical effort, a logic circuit delay model, in 1999. It is such a great model that I teach it in my class.

He also published a paper in 2003 on the use of electric coupling for chip interconnection. That was about the time when my team started our research on using magnetic coupling for chip interconnection. When I was MacKay Professor at UC Berkeley in 2007, I had the honor of joining him in faculty meetings.

Back to the subject at hand. Asynchronous circuits use dual-rail logic where the two outputs remain equal while computation is in progress. When one output switches, a signal indicating that computation is complete is transmitted to the circuit in the next stage together with the result of the computation.

While asynchronous design consumes more transistors and interconnects than its synchronous counterpart, there may come a time when you end up with a gain given the waste in synchronous design.

I thought the time would come in the 7nm generation. However, **because FinFET, which was introduced to improve transistor gate control, has better performance than expected, asynchronous design has not been able to reach a crossover**. Since it looks like transistor structure innovation will continue, it may be a while before asynchronous design gains adoption.

Nevertheless, asynchronous design is well suited for parallel data processing using wired logic as exemplified by neural network, which has attracted a lot of attention through its adoption by AI.

(But then we have had numerous instances of wavering in the past and sometimes settled on the wrong prediction).

Rhythms in Nature

One day in 1665, Christiaan Huygens, who proposed the Huygens–Fresnel principle based on the wave theory of light together with Augustin-Jean Fresnel, noticed accidentally that the pendulums of two clocks placed side-by-side on the same wall of the room were moving in synchronization. One pendulum always swung to the right when the other to the left. Even when he intentionally decoupled their motions, they would eventually return to being synchronized.

However, when he put the two clocks on separate walls removed from each other, synchronization did not occur. Huygens postulated that the synchronization was the result of a very weak interaction between the two clocks.

In fact, rhythm can be found everywhere in the world. And when rhythm meets rhythm, they synchronize.

For example, when people walk across a suspension bridge, their steps can reinforce each other to cause the bridge to swing widely even without coordination. This is what happened to the Millennium Bridge across the Thames River in London in 2000. Things like trends and traffic jams are also rooted in the phenomenon of synchronization.

Synchronization also occurs in insects and cells. In southeast Asia, many fireflies are found gathering in a mangrove forest and emitting blinking lights in synchronization.

In mammals, the suprachiasmatic nucleus in the hypothalamus of the brain has about 20,000 cells which synchronize to create a biological clock inside the body to generate the rhythm for things such as the sleep cycle. Inside the heart, there are about 10,000 pacemaker cells which tirelessly fire in synchronization to generate a heartbeat for a total of 3 billion times in a lifetime.

Even heartless, inanimate objects synchronize. Superconductivity is achieved when many electrons march in lockstep resulting in almost zero electrical resistivity. Meanwhile, the formation of laser into a high energy beam is the result of many atoms emitting photons with synchronized phase and frequency.

On the other hand, we can always see what appears to be a rabbit (to some Asian cultures) on the moon in the night sky because the moon's rotation is synchronized with its revolution around the Earth such that it always faces the Earth with the same side. Furthermore, the gravitational pull of planets in the solar system can synchronize and be aligned to send meteorites from the asteroid belt towards the Earth, which may have caused the extinction of dinosaurs.

Synchronization also occurs in man-made networks and virtual space. Power generators connected to a high voltage power grid can synchronize naturally. It is the consequence of speed alignment through energy transfer from generators with high rotational speed to those with low rotational speed. This can lead to chain reactions resulting in disruptions when anomalies occur. In addition, routers on the internet have been observed before to synchronize like fireflies causing sudden fluctuations in traffic.

The first engineering attempt to control synchronization was made by Robert Adler in 1978 when he authored an analysis of the frequency entrainment phenomenon of oscillator circuits.

My research team was probably the first to attempt to synchronize more than two circuits through coupling. We succeeded in 2006 in using transmission lines to couple and synchronize the outputs of four oscillators integrated on the same chip. Next, in 2010, we discovered the group synchronization phenomenon resulting from magnetic coupling of four chips stacked in 3D and used it to develop a technology for precise clock distribution to the chips.

Nonlinear analysis continues to be applied to understand such group synchronization phenomenon.

4.5   [Column] A Model of Group Synchronization

The physical laws that govern the thermodynamic behavior of the macroscopic world, which humans can perceive, are the law of conservation of energy (the first law of thermodynamics) and the law of increasing entropy (the second law of thermodynamics).

In contrast to the law of conservation of energy, which states that the total quantity of energy is conserved in its entirety while changing its form, the law of increasing entropy describes the supplemental fact that the quality of energy (in terms of whether it can be effectively used by human beings) degrades irreversibly.

It was Ludwig Boltzmann, a pioneer in statistical mechanics, who explained the concept of entropy, first introduced by Rudolf Clausius of Germany in 1865, from the microscopic standpoint of atoms and molecules. The law of increasing entropy refers to the fact that the disorder of a microstate can only increase irreversibly.

In this way, the world seems to be going from motion to stillness, from being structured to structureless, from life to death, and eventually, as postulated by some including

Boltzmann, "the heat death of the universe" may come. Does the dissipation of energy mean the gradual collapse of the universe?

No, I don't think so. Around 1967 when the Belgian chemist Ilya Prigogine formulated the concept of dissipative structure, efforts were started to understand the mechanism of order where structure (creation) emerges from the constant dissipation (destruction) of energy.

In other words, the forces that drive the creation of entropy and the disappearance of structure and motion are simultaneously the driving forces that create structure and motion.

For example, when a candle burns, it dissipates heat to the surroundings, creating a structure in the surrounding space in the form of a thermal gradient. This structure collapses with the dissipation of heat, i.e., the increase of entropy. The dissipated heat will certainly not spontaneously create a structure. On the other hand, the continuous discharge of entropy into the air generated by the combustion creates a structure in the form of a flame.

After the flame is put out and the generation of entropy ended, thermal equilibrium is reached. However, the candle retains its form, and what is described as the heat death of the universe does not immediately follow.

To prevent heat flow, the candle and its surroundings are given a limited amount of energy. After entropy reaches its maximum within these constraints, the atoms and molecules within each material become sluggish, and a metastable state is reached where individual shapes and properties are preserved. When the candle is lit again, a flame is recreated together with the generation of entropy.

Does the entropy released with the flame stay on Earth? The Earth receives energy primarily from sunlight, and releases entropy and energy into space via infrared radiation. In addition, the temperature differences between the sky, ground surface, and deep underground drive the efficient circulation of energy and release of entropy.

Our study of physics and engineering in university deals with linear systems. The defining feature of linear systems is that the overall characteristics of the system can be understood from the linear combination, or simple synthesis, of the properties of its components. In other words, the sum of its parts is the very whole itself.

That is why even very large and complex problems can be solved by breaking them into small pieces, solving them individually, and combining the answers to create the whole picture. For example, a gas in which the molecules move completely independently can be analyzed using linear statistical mechanics.

However, that is not the case for solids and liquids because of the strong interactions between their molecules. But that would make the problems too complicated to solve, so we apply linear approximation. Within a small range of changes around an operating point, we can treat solids and liquids like gases and infer their properties by analyzing them as gases.

In many cases, matter is a nonlinear aggregate of microscopic elements where interactions between elements cannot be ignored. As a result, even if it is a linear system when the inputs are small, it becomes nonlinear as the inputs increase.

For example, the multiplication of bacteria will stop when they become so numerous as to deplete all available nutrients. On the other hand, the output of an amplifier circuit will not remain proportional to its input beyond the supply voltage.

As we learned in wireless engineering, when a sinusoidal wave is input to a nonlinear system, new harmonics and waves from intermodulation appear at the output. Alternatively, as when water freezes or metal exhibits superconductivity, phase transition phenomena can occur suddenly, where new spatial order emerges, or the properties of matter undergo significant changes. These phenomena, which give rise to something new, result from close interactions between the constituent elements.

What is indispensable for the understanding of such complex nonlinear phenomena is the brute force approach of computer simulation and the insight from mathematical modeling that strips away everything but the essentials.

How does a group of nonlinear elements synchronize? How can a group synchronize without a conductor or cues from the environment?

Initially, the study of synchronous phenomena was conducted independently by biologists, sociologists, physicists, mathematicians, astronomers, and engineers in their respective research fields. Eventually, building upon such research results, the science of synchronization converged into the study of coupled oscillators and developed into a nonlinear science in which many coupled oscillators interact with each other. Let us explore the mechanism of the synchronization phenomenon (entrainment) by looking back at how great mathematicians and physicists have tackled this difficult problem.

Norbert Wiener, a mathematician and cybernetics advocate at MIT in the United States, was the first person to work on explaining group synchronization.

His intuition was that alpha waves serve as the master clock of the brain. He hypothesized that the phenomenon of frequency entrainment results from the coupling of neurons with disparate rhythms.

However, he died in 1964 without proving it. The following year, a student at Cornell University in the U.S. discovered a mathematical approach to solving the problem. That student was theoretical biologist Arthur Winfrey.

Before introducing Winfrey's work, let me explain the cardiac pacemaker model proposed by Charles Peskin, an applied mathematician at New York University, USA, in 1975.

Peskin analyzed the oscillation of potential in the cell membrane of a pacemaker cell with an oscillator circuit. In this nonlinear model, the cell membrane (capacitor) is charged via its leakage channel (resistor) and fires and discharges as soon as its potential exceeds a certain threshold.

In Peskin's model, the coupled oscillators are of equal strengths and interact only at the instant of firing. That is, when one oscillator fires, it immediately causes the potential of the other oscillators to jump by a certain amount. If this causes a cell to exceed the threshold, that cell will also fire immediately. However, mathematics at the time could not handle such a large oscillating system where internal interaction is caused by such impulses. Therefore, Peskin limited his work to the case of two weakly coupled, identical oscillators and proved that the oscillators are always synchronized.

Now, Winfrey created a mathematical model of the coupled oscillators that further abstracts from Peskin's model and retains only the essentials. It is generally the following model.

Consider first two oscillators rotating in a circle in the same direction and at the same speed. A pulling or repelling force acts between the two oscillators. This interaction is a nonlinear force determined by the positions, or phases, of the two oscillators. The interaction causes the two oscillators to change speed and eventually synchronize either in or out of phase. Specifically, if the interaction is a pulling force, the two oscillators will become in phase; otherwise, they will become 180 degrees out of phase.

If the original speeds of the two oscillators were different, they would stabilize to close to a phase difference of 0 or 180°. The deviation is determined by the magnitude of the interaction. The stronger the coupling is, the smaller the deviation. And if the coupling is smaller than a certain value, no synchronization will occur.

Next, for a larger number of oscillators, an equation is formulated to describe the relationship of the speed at which each oscillator moves relative to the group. At any point in time, the speed of an oscillator is determined by three factors. The first is the intrinsic frequency of the oscillator. The second is its sensitivity to the totality of external influences. This is determined by the position of the particular oscillator. The third is the gross influence exerted by all other oscillators. This is determined by the positions of these other oscillators.

The equation is actually a set of nonlinear simultaneous equations. Although they cannot be solved analytically, it is possible to estimate the behavior of the group in simulation. Starting with certain initial positions of the oscillators, the equations calculate the instantaneous speeds of the oscillators and their positions at the next instant. By repeating this calculation, the behavior of the oscillators can be predicted.

As Winfrey repeated the simulation with different combinations of sensitivity and coupling coefficients, he made several observations. For example, there are both cases of spontaneous synchronization and desynchronization. Moreover, in the case of spontaneous synchronization, there is no irreplaceable oscillator that is at the center of the process.

And the most significant finding is that as the group becomes increasingly homogeneous, it will suddenly synchronize after reaching a certain critical point. It is just like when water is cooled to the freezing point, it undergoes a sudden phase transition and turns into ice.

A sudden phase transition occurs when the relative strengths between the force that seeks to create order and the force that seeks to destroy it is reversed. If we consider the width of the frequency distribution as temperature, the oscillators corresponding to water molecules "solidify," or synchronize, through coupling to create a macroscopic temporal order.

Winfrey thus crossed an important bridge between the two disciplines of nonlinear dynamics and statistical mechanics. He published *The Geometry of Biological Time* in 1980.

Winfrey's accomplishment opened the door to a new academic discipline, which saw the emergence of talented scientists one after another.

Dr. Yoshiki Kuramoto, a physics professor at Kyoto University, proposed the Kuramoto model, which is an improved version of the Winfrey model, and obtained an analytical rather than simulated solution. Steven Strogatz, a professor of applied mathematics at Cornell University, has worked on explaining the pulse coupling between biological oscillators and co-authored the theory of "small-world" networks with his student Duncan Watts, which has advanced nonlinear science into the realm of social networks.

Democratization: More People

5

5.1 Time Performance: Time Being Money

Cost Performance and Time Performance

One often hears the expression "The cost performance is good." Cost performance is the most important metric in the semiconductor industry.

However, recently **time performance** has become an important metric as well. There are two reasons for that.

The first reason is because society is changing from being capital-intensive to knowledge-intensive.

In the recovery after the war, Japan strove to become an industrial nation, and furthermore, an electronic nation through the development of semiconductor technology. Both the industrial society (Society 3.0) and information society (Society 4.0) are capital-intensive societies. Bigger is better, and mass production of standardized products and mass consumption are encouraged. But it has become clear that the resultant increase in the burden to the environment is limiting growth.

Japan is a rapidly aging society with low birth rate. The new Society 5.0 we are aiming for is a human-centric society. In other words, it is a society where everyone is expected to share their wisdom and knowledge.

A society where wisdom gives birth to value is also one which makes use of the individuals. It is Japan's new strategy to create a society with sustainable growth where everyone thrives.

Digital innovation is the driving force towards that goal.

Unexpectedly, the spread of Covid-19 is accelerating digital innovation. **Digital innovation starts with the creation of a platform, where speed is the essence**.

In a capital-intensive society, materials are the resource and things deliver the value. Specifically, materials are used to make components, which are used to create products.

T. Kuroda, *The Super-Evolution of Semiconductors*, Synthesis Lectures on Engineering, Science, and Technology, https://doi.org/10.1007/978-3-031-60518-5_5

Wisdom in the form of services, design, and market strategy and so forth is then added to deliver societal impact. In this scenario semiconductors are components. Components must be low cost.

On the other hand, in a knowledge-intensive society, data is the resource, and wisdom delivers the value. Specifically, AI is used to analyze data collected using IoT and 5G, and the results are used to create services and solutions. The power of semiconductors is then added to deliver societal impact.

In other words, there is a reversal of roles in value creation, where semiconductors are shifting into a role of higher value. **The semiconductor business must also transition from the component business of the past to a societal impact business. A new strategy is required.**

Another reason for the heightened importance of time performance is because semiconductors are transitioning from being the staple food of industry to part of the infrastructure of society.

In a capital-intensive society, the infrastructure consists of roads, harbors and railways which transport materials being used as the resource. In a knowledge-intensive society, however, information networks which move data as the resource form the infrastructure. And information networks are realized using semiconductors.

Cost performance is important for the semiconductor business with semiconductors being components. Because consumer products like TV, PC, and smartphone are replaced once every several years, devices with high cost performance that come later to the market are bought by consumers to replace their older predecessors. Therefore, cost performance is important.

By contrast, because the replacement cycle for industrial products such as communication equipment and robots is more than 10 years, even if higher cost performance products are available later on, businesses are not motivated to purchase them. In the end, products that get wide adoption are those which are first to market.

As a result, **time performance is important for the semiconductor business in Society 5.0. "Time is money." Time is determined by development efficiency, while performance is determined by power efficiency**.

Semiconductors Required for Post-5G

With 5G, base stations use software virtualization to deliver diverse services and address various use cases. In other words, it is necessary to construct a flexible network which implements functional virtualization of general-purpose servers and slicing.

On the other hand, beyond 5G, because the signal cannot travel far, the area covered by a cell shrinks, and as a result, base stations are being miniaturized. In other words, base station power, volume (size), and weight must be reduced so that many of them can be inexpensively deployed in urban areas. Specifically, telecommunication carriers are asking for "**5W, 5L, and 5kg**" as target.

Table 5.1 5G Base station hardware time performance comparison

	Server	FPGA	ASIC	Agile 3D-FPGA	Agile 3D-ASIC
Development time	0	6 months	14 months	1 month	6 months
Development cost	0	$10 M	$45 M	$2 M	$15 M
Production cost (100 k units)	$500 M	$200 M	$4 M	$250 M	$5 M
Power	50 W	30 W	6 W	15 W	3 W
Volume (Size)	3 L	2 L	1 L	1 L	0.5 L
Weight	10 kg	1 kg	0.04 kg	0.5 kg	0.01 kg

RaaS is undertaking the R&D of Agile 3D-FPGA and Agile 3D-ASIC

The limited power consumption required of small base stations limits the performance of their servers. To compensate for the inadequate performance, hardware accelerators with high power efficiency have become necessary. The trend is to add FPGA- and ASIC-based network cards to the servers so they can handle the computationally intensive routine processing in hardware.

As a result, even though general-purpose servers are adopted starting with 5G (specialized hardware using ASICs was used up until 4G), **what determines performance and cost is FPGA and ASIC**.

The added costs, power, volume, and weight when FPGA or ASIC accelerators are added to general-purpose servers have been computed and compared in the following Table 5.1. While the absolute values vary with the assumptions, the table can be used for relative comparison.

If you compare realizable performance under the same power constraints between server, FPGA, and ASIC, the ratio is 1/50:1/30:1/6, or approximately 1:2:8. In other words, ASIC is a very effective way of delivering performance. CPU and FPGA have poor power efficiency because it is necessary for them to include a significant number of extra circuits in order to support programmability. The need to support software backward compatibility creates an additional burden by embedding upgrade history in the circuits.

However, ASIC has its challenge in high cost due to limited production volume. From 7 nm process onwards, mask cost alone amounts to $10 M, and EUV lithography will remain expensive until its equipment is fully depreciated. Nevertheless, if production volume reaches 100 k units, its total cost including development cost can be reduced to 1/10 of server. It can thus increase server's profitability proportionally.

The recent **worldwide trend of developing specialized chips (ASICs)** instead of using general-purpose chips **is driven by the goal of achieving lower power and cost**. In other words, it is because of better cost performance. By using ASICs, you can achieve both higher performance and lower cost.

There was a time when communication equipment makers also actively developed ASICs. In the 1990s, transistor count was only on the order of 100 k, and an ASIC could be developed in a few months. By contrast, transistor count has climbed to a few billion, and design alone can take more than a year.

In other words, the challenge of ASIC is that as device density increases, the time required for design and verification has increased to an unacceptable level. On top of that, Japan has the challenge of continuously losing its ASIC design capability. The outflow and loss of engineering talent resulting from the decline of Japan's semiconductor industry has been painful.

Because communication is an infrastructure business, what is most important is business continuity. Telecommunication carriers who can secure vested interest in the form of frequency allocation and create a stable business have the ability to decide a specification and attract bidding by multiple suppliers. Meanwhile, as a result of M&A driven by intense international competition, only a few mega suppliers have been able to survive. Nevertheless, the recent trend of securing supply chain to ensure economic security is driving a re-examination of such an industry structure.

In the communication equipment business, it is common for the supplier that is first to market to win market share. As a result, supplier competition has led to the shortening of the lead time from when specifications are finalized to when products are launched.

Time performance is important for AI too. That is because AI technology is advancing so rapidly that technology from a few years back is already obsolete.

Making Full Use of the Computer

I have heard the following from someone in the telecommunication carrier business. "In contrast to Chinese makers taking only 2 months to design an FPGA, it takes Japanese makers more than 6 months. While it may partly be due to differences in business culture, close examination reveals that Chinese makers achieve the shorter turnaround time by throwing a lot of human resources at the task."

The strategy that Japan should adopt is one that drives the task with computer and leaves **no human in the loop**, instead of relying on vast human resources.

At RaaS, we are pursuing large **time performance** by engaging in R&D to achieve a tenfold increase in development efficiency as well as tenfold increase in energy efficiency.

To achieve a tenfold increase in development efficiency, we are pursuing R&D of an **Agile Design Platform (Agile 3D-FPGA and Agile 3D-ASIC** in Table 5.1) and adopting open architectures such as RISC-V through international collaboration. Our goal is to use the computer to fully automate design and verification and eliminate the possibility of errors by taking humans out of the loop.

In parallel, to achieve a tenfold increase in energy efficiency, we are pursuing R&D of 3D integration technology as well as utilizing advanced CMOS processes through our alliance with TSMC. By stacking multiple chips and integrating them into a single package, we will be able to shorten the distance of data movement by orders of magnitude, thereby significantly improving energy efficiency.

In fact, this strategy shares many objectives with the Defense Advanced Research Projects Agency (DARPA)'s Electronics Resurgence Initiative (ERI) in the US. The difference is the incorporation of Japan's area of expertise of 3D integration. The Agile Design Platform will be the product of combining EDA and 3D integration.

Japan's telecommunication carriers outsource their chip design to Qualcomm (US), Mediatek (Taiwan), Broadcom (US), and HiSilicon (China). Our goal is to make it possible for them, as chip users, to use the computer to design their own advanced chips without relying on overseas chip design houses.

5.2   Agile Development: Chip Development Methods in the AI Era

From Waterfall to Agile Development Model

The mainstream model for system and software development has been a waterfall model. It starts with writing specifications and development plans, followed by top-down design and implementation based on the plans. Because development proceeds sequentially where you are not supposed to go back to preceding steps, it fits the analogy of a waterfall where water flows only downwards and never returns from downstream to upstream.

Agile development is the opposite. It is a bottom-up approach where the design is broken into small units each of which is developed through iterations of implementation and test. The model first appeared in 2001. Since it generally results in shorter development time compared to the waterfall model, it is called agile development for being quick and adaptable.

Another advantage of agile development is the ability to change or add to the specifications midway through the development. However, because of that, it is easy to deviate from the original direction of development, and it is difficult to see the big picture and manage the schedule, which are its shortcomings.

Given that changes will likely occur to the specifications and design midway through development, it suffices to start with just rough instead of detailed specifications. By establishing the resilience to flexibly adjust to changes as they occur during development, you gain the ability to better address customer needs.

Once the rough specifications and plans are decided, the system is broken into small units. While development proceeds through planning, design, implementation, and test, releases are made iteratively once every 1–4 weeks.

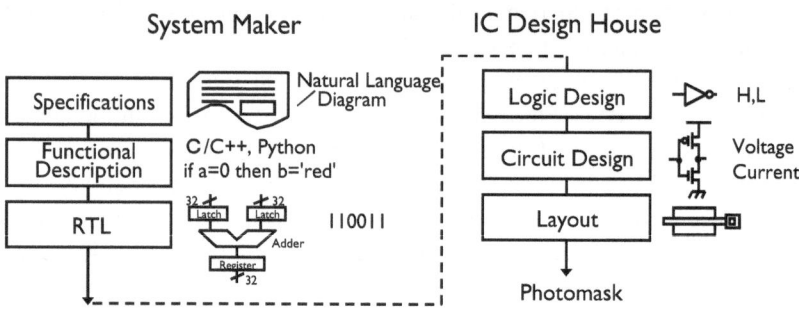

Fig. 5.1 System makers develop chips in an agile manner like how software is written

Chip design, however, is top-down.

Specifications written in words and diagrams are translated into hardware description language such as Verilog, while clock cycle-based processing sequences are coded into RTL (register transfer level). This is followed by logic design, circuit design, and layout, which is finally made into geometric patterns in photomasks. In this manner, chip design is completed as a series of conversion steps with decreasing level of abstraction (Fig. 5.1). This conversion can be automated with chatGPT.

To improve design productivity, computer automation was introduced into the design process in reverse order starting from downstream which involves a lot more information. It was introduced to mask design in the 1970s and layout design in the 1980s. Logic design was automated in the 1990s. High-level synthesis that automates system design entered into research stage around 1990, and partial, practical application around 2010.

However, the common way to improve system design productivity is the reuse of RTL. Design IPs that implement general-purpose functionalities such as processor core and memory controller are widely available. Furthermore, RTL of specialized circuits are usually not created from scratch, but from the reuse of previously designed RTL.

Even with such practices, large-scale chips today, such as Apple's A12 processor which integrates 6.9 billion transistors, require several hundred engineers spending several years to develop, costing 100s of millions of dollars.

As integration density continues to rise exponentially, the conventional development approach is reaching its limits.

Add to that the emergence of AI. **AI is evolving rapidly. Even the technology from last year is now inferior. It simply is too risky to engage in development of chips that requires time measured in years and costs in the 100s of millions of dollars**.

Agile Development of Chips

We believe that agile development can be applied to system design and verification by chip users.

A system can be developed by dividing it into small units which are described using languages such as C/C++ or Python. The RTL of these units are automatically created using high-level synthesis tools, which are then assembled to generate the system from bottom up.

By enabling agile development of chips in a way similar to how software is written, development time and costs for system makers can be greatly reduced, and their risks significantly lowered.

Using high-level synthesis tools, RTL of various combinations of circuit performance and layout area can be created in an instant. Therefore, these tools enable search for the optimal RTL by trading off between performance and area, which is then implemented in FPGA or verified using ASIC simulation tools to support release iterations within short timeframes.

In conventional approach, after studying the specifications thoroughly the designer creates block diagrams and estimates meticulously performance of each block and congestion of signal connections before starting the design. However, since it is difficult to quantify performance and area in the early stage of the design, the designer must rely on their intuition and experience. More importantly, the task quickly becomes overwhelming as system complexity increases.

Using an agile development model, the system is divided into small units, the functional block of each is automatically created, verified, and released in iterations using the computer. ChatGPT can automate this process.

The bottom-up assembly of the released blocks can also be computer-automated. This is because by using high-level synthesis, it is possible to distribute control mechanisms into individual functional blocks, such that the overall control mechanism is established when the functional blocks are assembled together.

In other words, **large-scale chips can be created by assembling individual functional blocks, much like parallel distributed programs in software**.

Using C/C++ or Python instead of RTL description can reduce the number of lines of code by 100x. As a result, the effort and time required by designers such as to review and simulate can be reduced by orders of magnitude.

Since structure of the circuit in high-level description is expressed in terms of parameters, it enables a wider spectrum of implementation options, as well as a good grasp beforehand of realizable implementation ranges of functionality, performance, and interface protocol.

In addition, by coupling the verification model to the design description such that changes to the latter are automatically reflected in the former, it not only makes it easy to confirm the extent of impact of the changes, but also enables effective assembly of the verification environment in conjunction with the design. In other words, it enables agile development of both design and verification simultaneously.

Using this approach, functional blocks are connected using dedicated control circuits, making it possible to increase energy efficiency. In conventional approach where IPs are connected to the CPU bus to allow the CPU to perform centralized control, it is difficult to realize high performance in complex processing such as required for 5G (communication), H.265 (video compression) or WPA2 (encryption).

Furthermore, in conventional approach where the RTL is designed with the intention for reuse in other projects, the circuit tends to be designed to be higher performance than is needed. By contrast, when high-level synthesis is used, circuits can be automatically generated each time with performance and area optimized for the particular usage.

Divide and Conquer

The first thing I learned about CAD at UC Berkeley is "divide and conquer." The idea is that by dividing a problem into small problems of similar sizes, you can find the solution to problems of any complexity by assembling the solutions of the individual, small problems. A lot of computer algorithms are designed using this idea.

Breakdown of the problem, solving of the small problems, and assembly of the solutions are carried out in a recursive manner. The result is a dramatic reduction in computation time.

For instance, if we compare the computational complexity of sorting algorithms for an input size of n, quick sort which employs "divide and conquer" reduces complexity to $O(n\log_2 n)$ compared to bubble sort which has a complexity of $O(n^2)$ [38]. This corresponds to a reduction from 1,000,000 to 9,966 when n is 1000, representing a $100 \times$ reduction, or a 100-fold speed-up.

The AI Age calls for rapid iterations of trial and error. AI is used to analyze large volumes of data to allow it to build a model. Then the model needs to be implemented rapidly to enable additional data collection and analysis to improve the model. And the process is repeated. It is vital to be able to efficiently execute such trial and error.

It is imperative to create a chip development model suited for the AI Age that meets the conflicting constraints of staying agile while enabling large-scale design.

I have learned a lot from China both regarding agile development and data collection. That reminds me of a foreign student from China who often said to me, "Professor, you are painstakingly overprepared."

5.3 Silicon Compiler: Creating Chips like Writing Software

Silicon Compiler 1.0

A compiler is a piece of software that converts source code to object code. Because source code is written in a high-level language that resembles a natural language, it cannot be understood by the computer as-is. Therefore, a compiler is used to translate it into a machine language in the form of object code, which is binary code executable by the computer.

Similarly, software that converts hardware specification into a silicon chip is known as a silicon compiler. Software that converts from Verilog, a hardware description language, to GDS-II, a language describing mask fabrication, is one such example.

In 1979, Dave Johannsen from Caltech published a paper entitled "Bristle Blocks: A Silicon Compiler." (A bristle block is the building block of a toy that is brush-shaped to allow individual blocks to stick together in any position. Its appearance looks like a semiconductor.) Since it was the same year when Carver Mead and Lynn Conway wrote the textbook for VLSI design entitled "Introduction to VLSI Systems" (which by the way fascinated and drew many of us into the world of VLSI design), one can say that silicon compiler is a very natural idea for complex chip design.

Johannsen's adviser was Mead [39], who in 1982 foresaw the coming of an age when specialized chips are made using silicon compilers and foundries in his paper entitled "Silicon compilers and foundries will usher in user-designed VLSI."

Johannsen co-founded Silicon Compilers Inc. with Edmund Cheng in 1981. By using the GENESIS tool from the company, one could design a chip in one-fifth the normal time by choosing from the tool's menu. The tool was used by Digital Equipment Corporation (DEC) to develop its MicroVAX minicomputer. However, the company did not otherwise achieve much success and was eventually sold. Another company which developed a silicon compiler—Seattle Silicon Technology—was not successful either.

Even today silicon compiler has not achieved wide adoption. Why is that the case?

If a software program has a bug, it can be patched and fixed later. By contrast, if a piece of hardware has a bug, it must be fixed immediately. Furthermore, while the performance of a software program is expected to evolve in conjunction with the hardware, hardware is supposed to meet its performance specification before it can be shipped. As a result, **hardware is much more difficult to design than software, and its development carries higher risk.**

While one-click compilation is the expectation in the software world, it is a dream in the hardware world. Even for the major design tool companies like Cadence and Synopsys, the compiler tools they develop are for skilled designers only. **Being able to make chips the way software is written has remained a wild dream.**

Silicon Compiler 2.0

Recently, interest in silicon compilers is rising again. However, the reason is different than before.

Today chip design is about optimizing PPA—power, performance, area. Once, area and hence chip cost had the highest priority. Eventually, performance in terms of operating speed became important, and **now power has the highest priority**. That is because chip power has reached its upper limit, such that only those who can increase power efficiency can extract proportionally higher chip performance. In other words, chip performance is determined by power efficiency.

Compared to general-purpose chips which can do everything, specialized chips can achieve orders-of-magnitude higher power efficiency by eliminating unnecessary circuits. However, since specialized chips have much smaller production volumes than general-purpose chips, each chip bears a much larger portion of the development cost.

Chip design technology has not been able to keep pace with Moore's Law. As a result, development cost has been rising rapidly in recent years, amounting to something on the order of 100 million dollars. As an example, if the development cost is $100 million and 10 million chips are manufactured at a unit cost of $10, development cost will amount to half of the total chip cost. Consequently, if the development cost can be reduced to 1/10x, even if chip area is increased by 1.5x, there is still a 25% costdown.

In the past development cost was small enough that chip area had the highest priority. Since development cost has been rising rapidly, its reduction has become important. Furthermore, with the fast pace of technological changes nowadays, besides lowering cost, shortening development time is also critical in order to reduce risk.

Combining the use of ASIC for orders-of-magnitude reduction in power with the use of a compiler to reduce development cost and time, albeit at the expense of somewhat worse performance and area, can be a profitable formula. And if that leads to more chips being designed, it is possible to reduce mask cost from $10 to 0.1 million by using MPW (multi-project wafer) prototype pooling services.

By using high-level synthesis, chip description can be written in C, which will enable the number of chip designers to grow like software engineers. If the open source model can take root in the hardware world, its ecosystem can expand into a multitiered structure to enable mass collaboration. When that happens, it will become possible to make chips in the way software is written.

At d.lab we are engaged in the research and development of a design platform to implement high-level synthesis of Verilog from C, then use 3D-FPGA to build and verify the system, followed by compilation from Verilog to GDS-II for ASIC development.

Fig. 5.2 Making specialized chips the way software is written by using silicon compiler

Our goal is to democratize access to silicon technology. We want to enable system developers to rapidly create ASIC hardware. To achieve that, we aim to enable making chips the way software is written by using silicon compiler to improve development efficiency by tenfold (Fig. 5.2).

Renaissance

The year was 1986. I was at Toshiba exploring the possibility of working with Silicon Compilers Inc. Through that work I came to know Tom Ho, who would later become my best friend.

Tom graduated from UC Berkeley after immigrating from Macau to California. After serving as chief of the 80,286 processor design at Intel, he joined Silicon Compilers Inc. at the invitation of Edmund. When we met, he was 31 and I was 27 years of age.

In a San Jose motel, we discussed circuits while drawing circuit diagrams in a notepad. We were so absolved that we lost track of time. It was Tom who taught me that an inverter with its output short to its input makes the best SRAM sense amplifier. The ABC (automated bias control) circuit which I published later in 1991 was an idea that germinated out of that discussion.

I asked Tom where he learned about circuits and his reply was from Carlo Séquin at UC Berkeley. And when I said I wanted to visit Berkeley, he took me on a 1.5-h journey to get there, while carrying the bulky GENESIS manual along with him.

I was a visiting scholar at UC Berkeley in 1988. My host was Séquin who developed RISC-1 together with David Patterson [40]. It was at UC Berkeley where Donald Pederson developed the SPICE circuit simulator in the 1970s, and where Richard Newton, Alberto Sangiovanni-Vincentelli, Robert Brayton and others led the research on design technologies including automatic layout and logic synthesis in the 1980s. It was a prolific time when major companies like Cadence and Synopsys were born one after another. However, from around 2000, the EDA market gradually reached saturation, while technological advance stalled.

But recently, I hear a lot about students at UC Berkeley taping out repeatedly about once a month by writing RISC-V in Chisel. **I can sense the coming of an EDA renaissance**.

Tom, don't you want to do silicon compiler again!

5.4   Democratization of Semiconductors: Agile-X

Agile-X

What will be the next technology after electronics? Spintronics, photonics, or …

The Ministry of Education, Culture, Sports, Science and Technology (MEXT) of Japan has issued a call for applications for the "Initiative to Establish Next-generation Novel Integrated Circuits Centers (X-nics)" which aims to develop a new approach ("X") to research and development for the creation of energy-efficient, high-performance semiconductors and to foster development of professionals who will lead the semiconductor industry in the future.

We applied with the following proposal.

Whatever the next-generation technology is, it should be implemented quickly to deliver real benefits using semiconductor technology. In other words, agile is the new "X".

The proposal was accepted, and the "Agile-X Center for the Democratization of Innovative Semiconductor Technology" project was launched at the University of Tokyo in 2022.

The challenge with specialized chips is development efficiency. Even with 100 people working on the design, it can take up to a year. In addition, it takes four months to manufacture the chips. And the development cost can exceed $50 million. The development time and cost explode as the level of integration increases.

As a result, the number of companies capable of developing specialized chips in Japan is decreasing year by year and a hollowing out of the industry has begun. Even if a new semiconductor plant is built in Japan, it will not directly contribute to Japan's digital industry if it is only used by foreign companies. Japan needs to strengthen its design capabilities.

In the first place, taking a year and a half to develop the product itself is not consistent with the growth strategy for the digital age.

The importance of tightly integrating hardware and software has long been advocated by many industry leaders including Alan Kay and Steve Jobs.

However, achieving tight integration between the two is very difficult because software is continuously upgraded while hardware cannot be updated quickly.

Can you come up with a chip that can be updated like software?

I was once asked by a manager of a major electronics manufacturer with a straight face.

In other words, **we want to be able to design chips like writing programs and make prototype chips like compiling programs**. Agile-X aims to realize that dream.

If we can develop a platform that can reduce both the chip development time and cost to one-tenth, we can be sure that the population of developers who can design specialized chips will increase tenfold, and semiconductor chips will be democratized.

But is that really possible?

What is needed is a change of thinking. The industrial world is structured toward the mass production of standardized products. It is impossible to make it better than it is, but we can create something different. In other words, we can create a new industrial system oriented toward high-mix, low-volume production.

First, let's fully automate the design process by using the computer. We want "no human in the loop," meaning no one needs to be involved in the design cycle.

Naturally, the achievable performance will not be able to match that of something a skilled designer takes time to design.

However, following the 80-point doctrine from Toyota, an 80-point score is good enough if combined with some additional value-added.

It is more important to reduce the design time to one-tenth than to take the time to achieve a perfect score. In the digital age, time is money. Time performance is more important than cost performance. To that end, Japan has a lead in the high-level synthesis technology required.

Next, we can make prototypes quickly with semi-custom manufacturing.

This is similar to manufacturing semi-custom instead of tailor-made clothes. Specifically, standard transistors are made in advance, and wiring is used to customize them into specialized circuits.

If general-purpose transistors are good enough, and you need only 100 chips, you can purchase the necessary state-of-the-art transistors for one-tenth the price of a single wafer, in other words, approximately several thousand dollars.

And if all you need to do is wiring, the chips can be manufactured in a month. In the 1990s, gate arrays could be developed in a week. Although the number of wiring layers has increased significantly since then, it should still be possible to manufacture the chips in one month.

We would also like to develop a technology to draw patterns directly on wafers without manufacturing expensive masks. Although the manufacturing throughput will be lower, this is more economical for low-volume production.

In addition, chiplets can be used to enable reuse of design assets.

By combining all these approaches, it will not be just a dream to develop a specialized chip at one-tenth the time and cost. As a result, we can increase the population of designers developing specialized chips tenfold.

Currently, there are 10 times more software than semiconductor developers. They are developing systems using general-purpose chips, which consume a lot of power and cannot provide competitive services.

If Agile-X can develop a specialized chip in one-tenth the time and cost, software and semiconductor developers will be able to work together to integrate hardware and software and repeat the development and improvement process in rapid cycles. This is the kind of growth Agile-X aims to achieve in 10 years.

Contributing to the Advancement of Science

When Agile-X is applied to education, students can learn everything from systems to circuits and devices, in other words, from the beginning to the end of semiconductor development. Since students can experience both design and prototyping in just weeks, Agile-X can be used for gaining practical experience even in a class on a quarterly academic calendar.

If manufacturing equipment can be connected to the Internet so that clean rooms can be operated from across the country, even device manufacturing can be experienced in the metaverse.

Such an educational system is currently under development, which we hope to make available nationwide by FY2024.

If you think about it, it is researchers who constantly find themselves racing against time. Even if they are only one day behind their peers, their research will not be valued. At the same time, researchers need to analyze large amounts of data to uncover the hidden truth.

If Agile-X is applied to research, data can be analyzed quickly to explore the truth, or ideas can be proven and implemented to serve society. In other words, it can contribute to the advancement of science.

David Shaw is a scientist who has developed his own supercomputer and is working on drug discovery.

He earned his Ph.D. from Stanford University in 1980, taught computer science at Columbia University, and founded D. E. Shaw & Co. in 1988, growing it into one of the largest hedge fund firms in the world through asset management based on advanced mathematical modeling.

Later, when a family member developed cancer, he became interested in the molecular dynamics of proteins and founded D. E. Shaw Research, Inc. in 2001. The company brings together researchers in biology, chemistry, physics, mathematics, computer science, and engineering in a New York City skyscraper.

The properties of protein vary greatly depending on how the chains of amino acids are folded, which also changes how the protein reacts to drugs. To analyze the three-dimensional structure of a protein, 10,000 operations are performed for every single atom, and if the operations are repeated for 1 million atoms in milliseconds at femtosecond time increments, the number of operations will be 10^{22}. That is an enormous amount of calculation.

For that purpose, in 2009, a supercomputer specializing in molecular dynamics and consisting of 512 nodes was developed. Then in 2014, a new specialized chip was developed to increase its computing performance by a factor of five to 12.7 tera (10^{12}) operations per second. Appling this chip to 512 nodes, it became possible to perform 6.5 peta (10^{15}) operations per second. The time required to perform 10^{22} operations is reduced to 20 days, making it possible to solve practical problems in realistic processing time.

Animation of the simulation results shows that protein moves smoothly but occasionally makes sudden motions. It was eye-opening to see large-scale variations in protein structure being reproduced.

Conversely, scientific advances can accelerate the evolution of chips.

We are currently collaborating with molecular neurobiologists at MIT. Recent developments in this field have been spectacular.

The neural network widely used today is actually a 70-year-old model. We have found that replacing it with a state-of-the-art model can reduce the power consumption of AI processing to one-hundred-millionth of the current level.

Without going into details, the key is to use about 10 different nonlinear functions for the synapses. Each nonlinear function can be realized in circuits by using a look-up table stored in memory.

Implementing this latest model will require about 10 chips, even when manufactured with state-of-the-art process. In other words, a neural network that learns from state-of-the-art molecular neurophysiology can reduce the power required for AI processing by a factor of 100 million, creating demand for state-of-the-art semiconductors.

In this manner, science and semiconductors will continue to promote evolution in each other, or co-evolve.

SPICE in Your Head

There was once a time when specialized chips were in demand. It was in the 1980 and 90s.

When I graduated from university and joined Toshiba, my job was to run circuit simulation day after day.

The computer I used at the time was an S/370, an IBM mainframe computer. The card reader would make rattling sound as it read in a deck of punched cards. Then, after a long silence that lasted hours, the printer would suddenly spit out page after page of results while making a roaring sound. By stretching out the pages of paper and connecting the symbols on it with a colored pencil, I would finally be able to see the operating waveforms of the circuit.

The circuit simulator was SPICE (simulation program with integrated circuit emphasis) developed by Prof. Donald Pederson at the University of California, Berkeley.

When you use a tool, you get better at using it over time, and it becomes more fun to use.

SPICE eventually took up residence in my head and gave me the ability to visualize the operation of a circuit just by looking at the circuit diagram, even before I could simulate it on a computer. In a sense, SPICE was my precious private tutor.

The SPICE manual was my bible. So, I was overwhelmed when I finally stood in front of the real Cesar Tower depicted on its cover.

I studied at Berkeley in 1988. At that time, Berkeley was truly a crossroads where international semiconductor brains circulated. In the era of specialized chips, Berkeley exerted a force like a magnet that attracted the world's brains.

Synopsys and Cadence, leading design tools companies, were born in Berkeley at that time. In addition, a prototype system that is a predecessor to the iPad was being tested in Berkeley 15 years before Apple made iPad into a product.

Now, we are once again entering the era of specialized chips. We hope that Japanese universities will become the crossroads of international brain circulation this time around (Fig. 5.3).

As a side note, there was a time when I ran SPICE myself and enjoyed simulating circuits with the PC version of it even on airplanes. But eventually I stopped doing it myself and left it to my students. And when I started drinking wine and watching movies on airplanes, SPICE slipped away from my head. It is truly regrettable.

Fig. 5.3 Democratization of semiconductors (from TSMC Chairman Mark Liu's speech slide, courtesy of ISSCC)

5.5   [Column] International Brain Circulation

Saratoga is an upscale residential neighborhood adjacent to Silicon Valley. David, Amin, Sahar, and TT are sitting deep in a poolside sofa. All four of them were my students at the University of California, Berkeley.

It has been 15 years since they graduated. David was an international student from China. After graduation, he worked in Silicon Valley and is now an investor and operating as a bridge between the U.S. and China. He is the owner of the house.

Amin and Sahar were students from Iran. Amin is currently an associate professor at Stanford University. He is married to Sahar, a former classmate, and lives in downtown San Francisco.

TT was an international student from Taiwan. He is currently an associate professor at National Taiwan University. He is at Berkeley on a sabbatical.

We get into the story about two professors from Berkeley's Department of Computer Science each donating $25 million to the University. They have started a very successful sky computing company, which is an evolution of cloud computing.

Then Amin talks about a startup he started in Silicon Valley. He explains why more and more funding is being raised these days without an IPO (initial public offering). Eventually, the conversation turns to the global economy, political economy, and history of their home countries, China and Iran.

Then David starts talking about a little historical expedition he and I have done.

David's friend Li-Chun had a Japanese scroll in her collection. David consulted me about its meaning in the spring of the previous year. In turn, I consulted with Professor Shuichi Sakai, Director of the University of Tokyo Library System. Thanks to his arrangement, the Historiographical Institute of the University translated the scroll and conducted related research.

The following spring, in 2022, when David visited Japan, we were recalling this episode at a tempura restaurant, when the conversation took an unexpected turn: it led to Morris Chang, the founder of TSMC.

As it turns out, Li-Chun's mother was a direct descendant of the Ashikaga family, whose history is older than that of the Tokugawa family, the most influential shogunate family of Japan. Her great-great-grandfather was Nariaki Tokugawa, lord of the Mito Domain (now Ibaraki Prefecture). Her scroll was of Tokugawa family origin. Li-Chun married a scientist named Robert Wu in Berkeley, where she had studied as a foreign student.

Robert Wu went to China to study under the scientist Qian Xuesen, who was named the Rocket King by Mao Zedong. In fact, Qian Xuesen's wife was Li-Chun's relative. Robert Wu succeeded in producing bipolar transistors domestically at the Chinese Academy of Sciences and contributed greatly to the realization of China's first domestically produced computer. However, as the storm of the Cultural Revolution swept through China, he and Li-Chun returned to Silicon Valley.

It turns out that this Robert Wu was Morris Chan's middle school classmate in China.

Meanwhile, Ta-lin Hsu, a fellow Berkeley alum senior to David, helped invite Morris Chang to Taiwan at the request of the Taiwanese government …

David and Li-Chun's friendship crosses national boundaries, and each of them can be traced to Morris Chan through a complicated path.

The threads of history weave complex patterns. Gravitational force works within the network of brain circulation nurtured by the international community. In this case, the common threads are semiconductors, Berkeley, and above all, friendship.

◇◇◇◇◇◇◇

It was 8 am in the morning in the summer of that year. I was visiting the Second Members' Office Building of the House of Representatives. I was invited as a lecturer to a study session for the Liberal Democratic Party (LDP) members. Several of those who had served as ministers sat in the room. There, I presented a talk entitled "International Brain Circulation—the Semiconductor Butterfly Effect."

In addition, I shared the story behind the establishment of the international conference, the VLSI Symposium. It was in the early 1980s, when the trade friction between Japan and the U.S. over semiconductors was intensifying.

Semiconductors are cutting-edge science and technology. Among researchers, there is mutual respect for the achievements of each other, from which friendship is also born.

Shoji Tanaka, a professor at the University of Tokyo, thought it was important to build a bridge to the US in the research community. So he spoke to Dr. Walter Kosonocky of RCA Laboratories, then the center of the U.S. semiconductor community, about starting a joint conference.

Walter escaped from Kyiv, Ukraine, in 1944 and immigrated to the United States, where he became one of America's leading semiconductor researchers after a lot of hard work.

Eventually, the first conference was held in 1981. Its digest states,

> The United States and Japan are the two major centers of activity in the development of VLSI technology. The objectives of these efforts are to achieve very high speed and very high density devices for high-quality information processing at low cost. We believe that the development of VLSI technology will initiate an Information Revolution affecting every individual of our industrial societies through improved productivity, new concepts in education and communication, and substantial changes in work and recreational lifestyle.

This is a truth that is still relevant today.

Eventually, Japan's semiconductor industry fell into decline due to exhaustion from the semiconductor trade war between Japan and the U.S. However, there are still many active researchers in Japan today who are known to the U.S. and the rest of the world. Thanks to the wisdom of our predecessors who established and managed this VLSI symposium, Japan has been able to preserve its world-caliber brainpower.

I attended my first VLSI symposium in San Diego in 1988. I was just 28 years old. At a banquet of over 500 people, my boss pointed to a man in the center of the crowd: "He is Walter Kosonocky, the key person in the U.S. Go say hello."

In fact, there was another young man in the audience who, like me, was attending his first VLSI symposium. He is Steve Kosonocky, Walter's son.

He and I are currently chairing the VLSI Symposium.

In closing, I sent the following message to the LDP members.

> Alan Turing, the English mathematician who created the principles of the computer, is often quoted as saying, 'Sometimes it's the people no one imagines anything of who do the things that no one can imagine.' **Such butterfly effect can be realized through international brain circulation**.
>
> Innovation comes from the collective brain. Therefore, in addition to the development of nanotechnology known as More Moore and More than Moore, it is important to democratize the technology of semiconductors, so as to achieve More People.
>
> It is also important to nurture a large number of highly skilled professionals who will serve as the core of this democratization, and to link them to the international brain circulation network. Universities are the crossroads and incubators of this network.
>
> No one knows what will succeed in research. Therefore, we must be prepared to shoot many wasted bullets. Regardless, international joint research will provide us with promising opportunities to connect to the international brain circulation network.

I would like to conclude my talk this morning by saying that people are the capital of Japan and are indispensable for the realization of new capitalism.

Epilogue: Super-Evolution

6

6.1 Massive Integration: Plenty of Room at the TOP

December 29, 1959, at the California Institute of Technology.

At the annual meeting of the American Physical Society, physicist Richard Feynman said,

> There's plenty of room at the bottom.

In other words, there were still many exciting things to be found in the microscopic world.

With these words of Feynman, the world began to explore microdevices.

Eventually, microelectronics was born, which then developed into nanoelectronics.

As semiconductor scaling approaches its limits, More Moore is an attempt to push it further. Japan has awakened from its long dormancy and is jumping right back into the competition starting at the 2 nm process.

On the other hand, research and development of More than Moore, which aims to create new value-added as alternatives to scaling, is also accelerating. Among the many current contenders, 3D integration technology is at the top of the list. We have entered the age of 3D integration of various devices in a single package.

With Feynman's words in mind, I assert that.

"There's plenty of room at the TOP."

While "bottom" is the exploration of nano (10^{-9}) dimensions, "top" is the exploration of massive integration in giga ($10^9 = 1$ billion) and even tera ($10^{12} = 1$ trillion) quantities [41].

© The Author(s), under exclusive license to Springer Nature Switzerland AG 2025
T. Kuroda, *The Super-Evolution of Semiconductors*, Synthesis Lectures on Engineering, Science, and Technology, https://doi.org/10.1007/978-3-031-60518-5_6

Chips integrating 100 billion transistors are close to becoming a reality. Intel CEO Pat Gelsinger predicted that by 2030 there will be one trillion transistors integrated in a single package. The age of tera-integration is close at hand.

In the 1980s, chips with 10,000 transistors were used in TVs and video players; in the 2000s, chips with 10 million transistors were used in PCs. Today, chips with 10 billion transistors are used in smartphones.

What will a trillion transistors realize?

In the future, semiconductors will create value by tightly integrating physical and virtual space. The stage for semiconductors will expand further, and the semiconductor industry will reach the 0.6% level of nominal GDP. In 2030, semiconductors are expected to be a $1 trillion market worldwide.

What exactly will semiconductor products that merge the physical and virtual space be used for?

For example, they will be used for automated driving.

First, they will begin to be used in limited places, such as on factory floors. Next, they will enable long lines of transportation vehicles to be driven unmanned on highways. In depopulated areas, they will provide a safe means of transportation for the elderly, who will have difficulty driving.

In long lines entering the parking lot, the vehicles would switch to fully automated operation so that people could get off first. This would eliminate congestion around the parking lot.

Even on ordinary roads, technology will become more sophisticated and gradually shift from driver assistance to automated driving. When that happens, urban traffic congestion will be eliminated. If drivers can have 20% more leeway in their driving, traffic jams may not occur in the first place. Cars and cities can work together to create that margin.

Robotics will also create a large market.

There will be a wide variety of robots, from those that assist with cleaning and nursing care to those that enjoy cooking and conversation.

Furthermore, sensing and other technologies can be used to reduce the need for manpower and create smarter functionalities, such as smart cities that automatically collect information from various parts of the city and smart factories that can automatically operate factories in a data-driven manner.

Semiconductors are the critically important core in all of this.

Japan, which is facing a declining birthrate and aging society earlier than anywhere else in the world, is truly a developed country with advanced challenges.

When the population increases rapidly, food shortage becomes a problem; when the population decreases rapidly, labor shortage becomes a problem. Semiconductors can help solve both of these problems; in particular, the problem of labor shortage can often be solved by adopting AI implemented with semiconductors.

We just need to have the right idea; in other words, innovation is required.

Innovation arises where ideas meet. More people is important so that more people can participate in creating innovation.

Compared to what was used in TVs and video players from 30 years ago, semiconductor integration has increased a million-fold.

In the past, the semiconductor industry was like a competition to build cookie-cutter houses: Multiple manufacturers competed to sell houses with the same specifications. The main selling point was price.

The semiconductor industry of the future will not be about building standardized houses, but about building communities. The kind of communities we want to build will be diverse.

And city planning cannot be done by just one company or industry. For example, automated driving requires collaboration between various industries, including car manufacturers, electronics manufacturers, telecommunications providers, data centers, public services, social infrastructure, insurance, and advertising.

Semiconductors will evolve from house building to community building.

6.2   Rich Forests: The Power of Ecosystems

Let's shift our perspective from semiconductor users to manufacturers. Japan is strong in semiconductor manufacturing equipment and materials.

Semiconductor manufacturing equipment is assembled from more than 100,000 parts, which is an extremely large and complex task since even a car has only about 30,000 parts.

Moreover, the specifications are different for each order. In other words, the list of over 100,000 parts is different for each unit. A custom production method is used where a small team of workers performs everything from product assembly to inspection. This is the ultimate in high-mix, low-volume production.

In addition, there are a wide variety of suppliers who deliver parts to semiconductor equipment manufacturers: There are suppliers who deliver custom-made parts to these suppliers, and those who deliver the materials used to make the parts. In other words, a huge network with multiple layers of hierarchy and a wide range of industries has been formed to support the operation.

Let's look at the specifics. Semiconductor manufacturing equipment performs a variety of processes, such as transporting and rotating wafers, and spraying liquid and blowing gas on them. Some processes are performed in air, while others in a vacuum. For this reason, piping for liquids and gases and wiring for power and control run throughout the equipment. They also perform a sewage function by collecting the treated liquids and gases. In our homes, water, gas, and electricity are supplied from the outside, consumed by various devices, and discharged into the sewage system. Semiconductor manufacturing equipment is the ultimate miniaturization of the lifelines of a house.

The electricity, liquids, and gases supplied to the equipment come from the user's factory, and the form of supply differs from user to user. In order to accommodate this, not only the way the equipment is made but also the upstream design differs according to the user's situation.

To create a complex system tailored to the user's situation, the supplier needs to be in constant conversation with the user so they can be in sync with each other and continue to support each other in taking on challenges in order to evolve together.

It is just like a forest ecosystem.

The same is true for materials used in semiconductors.

Semiconductor materials need to be carefully mixed and matched because the characteristics of semiconductor materials vary with each manufacturing equipment. In other words, semiconductor materials are also the ultimate in high-mix, low-volume production and can be described as tailor-made products.

For example, temperature is key in the manufacturing of semiconductor materials. The higher the manufacturing temperature, the faster the chemical reaction and the shorter the manufacturing time. At the same time, the higher the manufacturing temperature, the more variations there are in the material structure, and hence the lower the yield.

The manufacturer provides the user with recommended process conditions that allow the material to perform as expected. However, since users rarely handle materials under the same conditions, defects often occur. Therefore, to be competitive, the manufacturer needs to be able to use its database to rapidly optimize the process conditions for the user.

Therefore, a deep and vast network has also formed in the materials industry. Various materials, mechanical, and electrical suppliers go in and out of chemical plants, and they are supported by a multi-tier network underneath.

For another example, semiconductors require packaging, so suppliers provide special electronic materials such as reinforcing fibers, substrates, and encapsulants with low thermal expansion, which are necessary for packaging. In addition, electronic materials also require special processing using special chemicals such as surface treatment agents, reaction accelerators, and antioxidants. Therefore, materials manufacturers produce and sell specialized electronic grade materials separate from general grade materials.

I once heard a story from a manager of a materials manufacturer that he had considered but given up on the idea of transplanting such a Japanese industrial ecosystem to other countries. It is very difficult to uproot the network of small and medium-sized companies that support materials manufacturers and transplant them overseas.

We tend to focus on the big trees. The media focuses on the rise and fall of large corporations.

However, it is the rich soil that supports large companies, in other words, it is the ecosystem network which makes the industry strong. The ecosystem is why Japan is strong in semiconductor manufacturing equipment and materials.

TSMC does not build factories on undeveloped land (greenfields). They only build factories on brownfields, i.e., land with a rich industrial ecosystem. Kumamoto has that.

Japan's industrial ecosystem is rich. Even if a giant device manufacturer falls, the forest can regenerate if the soil is rich. Conversely, planting trees on infertile land will not produce a forest.

You can move a big company, like transplanting a giant tree, but it is not easy to move the soil that supports the giant tree. We must not fail to see the forest for the trees.

As I was thinking about this, I saw the NHK Special "Super-Evolution" on TV about the biological world (aired on November 6, 2022).

And then it all made sense.

6.3 Super-Evolution: The Mechanism that Nurtures Diversity

Those with a survival advantage survive and procreate. After all, the world is about competition. The weak will not survive.

It has been more than 160 years since Darwin's theory of evolution was formulated to explain the laws of nature. Cutting-edge science is now attempting to unravel the hidden evolutionary mechanism of living organisms.

It is one of cooperation, of helping each other, in addition to competition.

The Earth is covered with lush plants in vibrant colors. Plants account for 95.5% of the total weight of life on land (470 gigatons). There was a major turning point in the astonishing evolution of plants, which boast such an overwhelming volume.

Five hundred million years ago, the Earth was barren. Four hundred fifty million years ago, the ancestors of plants came ashore from the sea. After venturing into the frontier of land, plants spread to cover the land about 400 million years ago. However, at that time, plants did not yet possess a certain thing.

Let's revisit the discussion in Chapter I.

The birth of that certain thing during the Cretaceous Period (145 million to 66 million years ago), when dinosaurs roamed the Earth, triggered a major revolution that transformed the planet. It led to a dramatic increase in the number of terrestrial species.

It is also believed that prior to the Cretaceous Period, the number of species was about one-tenth of what it is today. However, after the Cretaceous Period, the number of species exploded.

What is that thing in plants that caused the great revolution?

It is flowers.

Using pollen in flowers, plants can now attract insects to come to them. In turn, insects carry and help spread the pollen. **In other words, a co-existence relationship is established between plants and insects.**

Until then, plants had served unilaterally as food for insects for 350 million years since they came ashore. However, with the birth of flowers, a major shift occurred in which insects were turned from being harmful to being helpful.

This triggered a rapid acceleration of evolution of living organisms.

When a plant evolved into a certain special shape, insects also evolved their shapes to match the change. Moreover, some plants used bright colors to compete for the attention of insects. Meanwhile, insects acquired the ability to fly to enable them to reach the flowers without failure.

This resulted in co-evolution where one side promoted the evolution of the other and vice versa.

Thus, forests became richer, mammals that fed on insects attracted to flowers diversified, and fruits produced from flowers helped primates evolve.

This is where my imagination runs wild.

Eventually, flowers acquired a new power.

It was a speedup of their life cycle. The time required from pollination to fertilization was reduced from a year to a few hours. This accelerated the evolution of all living organisms.

$$y = a(1 + r)^n$$

This is the formula for compound interest calculation, where r is the interest rate and n the number of times interest is compounded. Even if the principal a is small, its future value will become large if invested long enough.

Replacing n with 1/t yields the fundamental formula for the digital economy where t is the cycle time of development. This equation applies to both chip performance improvement and company growth.

In other words, repeating improvement over and over again in rapid cycles is a growth strategy for the digital economy. Rather than the rate of improvement (r), it is essential to increase the number of improvement cycles (n), i.e., shorten the development cycle time (t).

That is why being agile is the key.

Let's return to the discussion of co-evolution.

Mammals that fed on flowers and the insects that gathered around flowers diversified. In addition, the nutrient-rich fruits produced from flowers accelerated the evolution of our ancestors, the primates.

Without flowers, the Earth of today that is full of colors and life might not exist.

There is another world, also created by plants over hundreds of millions of years, which lies beneath the forest.

Fine threads of fungi known as mycelia are found attached to the tips of roots. Mycelia are found not only around but also inside roots, becoming integrated with roots and spreading throughout the soil. They are connected to and coexist with plants in the forest.

Plants obtain nutrients such as nitrogen and phosphorus through their roots. However, roots alone are not sufficient for that purpose. Most nutrients are absorbed by mycelia from the soil and passed on to plants.

In return, plants share the nutrients they created through photosynthesis with mycelia, creating an inseparable relationship.

Mycelia grow to at least tens of meters in length, allowing mycelia from different trees to connect with each other. As a result, trees throughout the forest are connected together.

This huge network of underground mycelia connecting trees together serves an amazing function.

Specifically, nutrients produced by photosynthesis are transported from one tree to another through their respective roots connected by mycelia. In this way, nutrients are also provided to trees that do not receive sufficient sunlight.

The mycelial network helps a tree that is about to die from a lack of photosynthesis by sending nutrients from a healthy tree. It is truly a network that helps the weak.

How does a small tree grow in the shade of big trees?

A newly emerged young tree in a forest of giant trees must endure decades or even centuries in the shade if unassisted. During that time, however, the young tree is able to obtain the nutrients they need to survive through the underground network. In this way, even in the shade of a dark forest, the tender young life is able to grow little by little.

In addition, nutrients are passed back and forth between evergreen and deciduous trees.

Deciduous trees, which are active in photosynthesis in the summer, share nutrients they created with nearby evergreen trees. In return, in autumn, evergreen trees send nutrients to deciduous trees that have shed their leaves. Through sharing, they help each other survive their respective harsh seasons.

It has long been thought that plants compete with neighboring plants for light and nutrients, in other words, they have been considered to be in competition. But in reality, that is not the case at all. Rather, they have created a stable ecosystem by building a strong cooperative relationship through a network.

The world of mutual support that extends beneath the forest is the way of life that plants, the king of the land that completely covers the Earth, have carefully preserved.

Helping each other is better than competing with each other in sustaining life. The Earth today is populated with such organisms.

Japan's semiconductor industry has been focusing on cultivating forests. It is now being recognized by the rest of the world.

It has an industrial ecosystem that adds co-existence and co-evolution to competition. This ecosystem, which has been fostered in Japanese culture and society, is about to breathe new life into the industry through international collaboration.

Japan has brought a new TSMC plant to Kumamoto Prefecture. As a result, various follow-on investments have been initiated. The emergence of a giant tree has revitalized the underground network that is Japan's materials and manufacturing equipment ecosystem which supports TSMC's manufacturing.

Furthermore, Rapidus was born out of Japan-U.S. collaboration. The search for new materials and the research and development of manufacturing equipment for the pursuit of the microscopic world are flourishing.

The U.S. and Taiwan excel at building large-scale systems by combining modules. On the other hand, Japan excels at the meticulous adjustment required to thoroughly respond to the diverse demands of its customers. Through international collaboration, we will be able to create synergy with our complementary strengths. A new era has begun in which Japan contributes to the world while evolving together with it.

The keywords are international collaboration, international brain circulation, networking, co-existence, and co-evolution.

We live in an age of 3D integration of heterogenous devices in a single package. What kind of forest are we going to create in the package?

6.4   Emergence of New Life: Connecting to the Next Generation

Shibuya in the morning of December 19, 2022.

I followed the signs through a complex network of underground passages that resembled an ant's nest to arrive at my destination, a high-rise building. From a conference room on the 23rd floor, I looked down at the city of Shibuya through the window. I used to know every corner of the city when I was a student, but now it has completely changed.

Fifty-five students gathered for the Second Semiconductor Circuit Design Hands-on Seminar co-sponsored by Google and d.lab at the University of Tokyo.

I asked a first-year doctoral student studying astrophysics why he participated.

> I use supercomputers to process vast amounts of data. We need as many additional high-performance computing resources as we can get.

Several sophomores who were completing the general liberal arts requirements at the Komaba campus of the University were also in attendance. I asked a female student the same question.

She replied, "When I move on to the Hongo campus to purse my major, I want to study quantum computing. So I wanted to use this opportunity to learn about classical computers first."

Students from the Graduate School of Medicine, the Faculty of Economics, and the Graduate School of Interdisciplinary Information Studies also applied to participate in the seminar.

In a traditional curriculum, students must first learn about electronic circuits, then semiconductor devices and integrated circuits. After that, they gain chip design experience during their fourth year of undergraduate or first year of graduate studies.

The students who gathered at the seminar, however, had no such prerequisite knowledge. What they brought with them was a strong sense of curiosity. Anyone who could use a PC and write programs could participate in the workshop. And if it sparked their interest, they could then pursue the study of electronic circuits and integrated circuits.

Let's make it a program in which students from technical colleges and high schools can also participate. Let's make it a contest and hold national competitions. Ideas are the key. Let's make semiconductors a tool that anyone can use.

Participants lined up to look at the chips on display. Their gaze was filled with curiosity. Confidence began to grow in me that we are on the right track.

What is the "flower" that will catalyze the super-evolution of semiconductors?

The journey to find it has begun …

6.5   [Column] the Secret of Imec's Strength

Imec is an independent international organization in semiconductors research. As much as TSMC is the leader in manufacturing in the horizontal specialization model of the industry, imec is the leader in research.

Headquartered in Leuven, Belgium, the company employs 5,000 researchers from over 90 countries and territories.

Imec is not a university and thus cannot award degrees. Nevertheless, imec has 800 doctoral students because it offers the most advanced prototype line and evaluation equipment in the world. In addition, imec covers living expenses, housing costs, medical insurance, and one-way travel expenses.

Technologies that have been tested and evaluated at imec are adopted around the world as something of guaranteed quality. That is why as many as 550 companies send their frontline researchers to imec together with huge sums of joint research funds.

Its total revenue is 420 million euros. Eighty percent of this revenue comes from foreign companies.

Participating companies are sponsors but they cannot participate in imec management. Imec executives regularly visit each company in turn to listen to their requirements and address them with flexibility and speed.

In other words, imec has an excellent research ecosystem to attract the world's brains and investments from companies. While maintaining neutrality, imec flexibly responds to customer requests and uses its talent pool of global personnel to quickly solve their problems. This attracts further investments, which are reinvested in people and equipment. In this way, a continuous cycle of growth is created.

It is a "research forest" that promotes innovation through co-existence and co-evolution.

So where does imec's strength come from?

To find the answer, we need to understand its home country of Belgium.

Belgium is a small country standing at the crossroads of Europe.

It has been repeatedly subject to domination and pressure from major powers such as Germany, France, Spain, Austria, and the Netherlands. And in Belgium itself, society is divided into two regions: the Dutch-speaking region of Flanders and the French-speaking region of Wallonia.

Under these circumstances, Belgium prefers to remain neutral. There are times when it shows reverence to its neighbors and times when it demonstrates undeterred shrewdness. The fact that the EU is headquartered in Brussels is probably due to the spirit of neutrality in Belgium. At the center of Belgium's academia is KU Leuven (KUL), founded in 1425 and said to be the oldest existing Catholic university in the world.

Roger von Oberstraten, who studied microelectronics at the University, became a professor at KUL in 1968 after earning his Ph.D. from Stanford University. In the following year he built the first clean room in Belgium.

He believed that no university could afford to have more than just a small clean room, and the only way to build a well-equipped clean room is if it is shared with other universities.

In 1982, the Flemish regional government created a comprehensive program to strengthen its microelectronics industry, which led to the establishment of the non-profit organization imec (Interuniversity Micro-electronics Centre) in 1984. The first director of imec was Professor von Oberstraten.

Imec's Good Governance Charter states that "All of IMEC's activities are based on one and the same mission: to precede by three to ten years, through scientific research, the industrial applications in the field of micro and nano electronics, nanotechnology, design methods and technologies for ICT systems."

That is how imec started as an inter-university microelectronics research institute, with a small team of about 70 people, mostly university researchers.

A turning point came for imec in 1999 when Professor Gilbert de Klerk succeeded Professor von Oberstraten as its second director.

He decided to turn its research into a business.

This major transformation was unwelcome by a lot of people around him.

However, imec took this opportunity to grow significantly.

Having studied at Sophia University, de Klerk had friendly relations with Japan and started collaboration with Japanese companies.

It was also fortunate that the Dutch company Philips is in Eindhoven, in the same Flanders region as imec, which allowed it to form partnerships with foreign companies that had done business with Philips.

ASML, a Dutch manufacturer of semiconductor photolithography equipment, was established in 1984 in the same year as imec. ASML was established as a joint venture between the semiconductor division of Philips (now NXP Semiconductors) and ASM International.

At the time of its launch, ASML was far behind Nikon and Canon in market share. While the Japanese companies pushed ahead with in-house development, ASML accelerated its development by collaborating with various manufacturers at imec. For example, ASML was able to obtain a lot of feedback in the early stages of its photolithography equipment development by having semiconductor device makers use the equipment on imec's prototype line.

By working with semiconductor device makers gathering at imec from around the world, ASML was able to quickly develop a user-friendly platform, which has led to ASML's success today.

The fact that Prof. Luc van den Hove, who took over as CEO of imec in 2009, specialized in lithography may have been an underlying factor in the successful development of ASML's EUV lithography equipment.

The reason for imec's success in winning worldwide support can be found in its policy of tying research and development directly to the user's business, which has become a common way of thinking today.

For example, in research, imec tends to build upon prior works. It does not try to be the first. Instead of aggressively striving to be the leader, imec builds up its capability while maintaining a friendly relationship with everyone and without making any enemies. Then, without fanfare, it got to the top of the world. It is the same strategy often used in business.

And the appeal of imec can be summed up in one word: customer-oriented.

To that end, imec has remained neutral. It has gradually reduced its reliance on government funding by attracting investments from around the world. In addition, it has further increased its neutrality and independence by keeping sponsor companies out of its operation.

It is flexible and accommodating in contract negotiations and selection of research themes. If a company makes a sincere request, it is willing to go above and beyond the rules and make various efforts to accommodate. That is how a trusting relationship is established.

Of course, priority is given to large, influential companies. Still, imec will not turn away smaller companies or companies seeking to rise to a dominant position in the market by working with them with flexibility. This is a sensible policy since there is always a chance that a novel technology will blossom in the future.

Why is imec working with d.lab?.

When asked by the Japanese media, imec's CEO Luc van den Hove responded, "Because d.lab has different ideas than we do."

This shows how much imec values diversity.

Yaw Debock, vice president responsible for promoting collaboration with academia, is an expert in magnetic materials. We once discussed the cultures of our two countries over udon noodles at a restaurant in a traditional Japanese-style house in Nezu, Tokyo.

To sum it up, the two peoples have many similarities.

Postscript

Semiconductor democracy and chip war are two sides of the same coin. In this book, my focus is on semiconductor democracy.

In the nineteenth century, Otto von Bismarck explained in a speech how iron made a nation. Indeed, iron created contemporary cities and gave birth to weaponry.

Today, the technology battleground is in semiconductors. Semiconductors make a nation. What will semiconductors create, and what will they break? It will be up to our imagination and wisdom.

Chip makers are engaged in fierce battles to dominate in the manufacturing of next-generation semiconductors.

However, semiconductors are evolving into such a giant collection of technologies that the challenges can no longer be overcome by a single corporation or even a single country. We really should think of semiconductors as a global commons.

As a result, we should not be instigating semiconductor wars; instead, we need to be building ecosystems.

Technology is getting more and more complex. Therefore, we must look at the entire forest instead of the individual trees. The challenge ahead for the world is how to nurture the forest, or in other words, create a rich industrial ecosystem.

We can take hints from the botanical world.

Nowadays plants flourish everywhere across the globe. The revolution that created the world of today was started by flower-bearing plants.

A co-existence relationship born between flowers and insects resulted in their co-evolution where one evolved by promoting evolution in the other, and vice versa.

The birth of flowers provided the trigger for the accelerated evolution of life forms.

According to Darwin's theory of evolution, only the fittest survive and are able to continue their family lines. In a nutshell, life is about competition. However, the hidden

T. Kuroda, *The Super-Evolution of Semiconductors*, Synthesis Lectures on Engineering, Science, and Technology, https://doi.org/10.1007/978-3-031-60518-5

mechanism of evolution that is being uncovered by the latest science is that living organisms not only compete with each other but also cooperate to help each other. That is the essence of the theory of super-evolution.

Similarly, to enrich the "semiconductor forest," it is important to find its "flower." It was with that thought in mind that I set forth my theory of the super-evolution of semiconductors in this book.

I started by describing how high-performance semiconductors are developed from the standpoints of More Moore and More than Moore.

Next, I shared my thoughts on what high-performance semiconductors can create from the standpoint of innovation, or in other words, the standpoint of More People.

Will we be able to find the "flower" that is needed to evolve semiconductors from the age of competition to the age of co-existence and co-evolution? It is going to take more than capital and Moore's Law to make semiconductors a global commons.

The need is to get many people involved. That is More People.

I would like to thank Yusuke Horiguchi of Nikkei Business Publications, Inc. for his exceptional assistance in the publication of this book. I would also like to thank my colleague Shogo Kondo for his help in revising the manuscript. I would like to express my sincere gratitude to him.

Tadahiro Kuroda
March 2023

For Further Reading

1. **Prologue: The Return of Spring**

(1) Shift to a horizontal specialization business model.

Vertical integration is a business model in which a single company integrates all processes from upstream to downstream, from product development to production and sales. Horizontal specialization, on the other hand, is a business model in which the core of the product is made in-house while other parts are outsourced. Which model is superior?

The semiconductor business is inherently vertically integrated. This is because comprehensive optimization of design and manufacturing is required. In the 1980s, however, specialized logic semiconductor ASICs were born, and the development of EDA (Electronic Design Automation; automatic design tools) and PDK (Process Design Kit; manufacturing technology models) as interface between design and manufacturing enabled manufacturing outsourcing.

TSMC, a specialized foundry for contract manufacturing, was founded in 1987 in response to the growing capital required to build factories. It earned the trust of its customers by not competing with them through its pure-play model, and attributes its success to the fact that "a company's success comes from being in the right place at the right time, with the right business model" (Wall Street Journal, June 19, 2021).

General-purpose semiconductors such as memory chips are still vertically integrated. Logic chips have also become more difficult to scale in recent years, and as transistor structures undergo major changes to FinFET and GAA, there is a strong need for design-technology co-optimization (DTCO). As an analogy, if it is no longer possible to freely draw a picture (design) on a white canvas (factory), the canvas must be modified to match the picture. Therefore, the construction of a production line becomes a joint effort with the customer. But if only a few large customers can do this, operating the factory becomes risky. Tuning of the business model is continuously required.

T. Kuroda, *The Super-Evolution of Semiconductors*, Synthesis Lectures on Engineering, Science, and Technology, https://doi.org/10.1007/978-3-031-60518-5

(2) Semiconductors, needless to say, are the key technology behind digitization, decarbonization, and economic security.

Since semiconductors possess material properties that lie between conductors and insulators, it is possible to use them to regulate current flow. In 1947, the transistor was invented. The name "transistor" comes from its ability to *trans*fer re*sist*ance between its input and output. It can be used for signal amplification through fine-tuned control, or as a switch by utilizing its on/off states. A logic chip is a large-scale integration of semiconductor switches.

Before the advent of semiconductors, vacuum tubes were used in electronic circuits. Vacuum tubes operate by controlling the flow of electrons which are emitted into space by heated electrodes. As a result, the electrodes grow thin over time and eventually break like a light bulb. By contrast, semiconductors are more durable because the flow of electrons is controlled in a solid state without heating. Therefore, chips are solid-state circuits.

The manufacturing of semiconductor chips involves many companies and takes months. The cause of the semiconductor shortage is similar to a car not being able to start and stop abruptly in a traffic jam. A production line operating at full capacity cannot immediately respond to changes in demand.

Semiconductors are used in a wide range of products that consume electricity. For example, even a lack of inexpensive semiconductors for wiper control in automobiles will prevent a car from rolling out of the factory. A disruption in the supply of semiconductors will have a major impact on the economy. This is one of the reasons why semiconductors are considered a strategic component for economic security.

(3) In order to encourage such investments, which will revitalize local regions, the recently enacted supplementary budget provides for 1.3 trillion yen (about $9.3 billion).

Competition in investment in the semiconductor industry is heating up. In the U.S., legislation has been passed to provide $52.7 billion dollars in subsidies over five years. In response, the Chinese government has announced measures amounting to more than 1 trillion yuan over five years. The EU too has announced a bill to invest 43 billion euros by 2030. In Japan, the Ministry of Economy, Trade, and Industry (METI) has allocated a supplementary budget of 770 billion yen (about $5.5 billion) in FY2021 and about 1.3 trillion yen (about $9.3 billion) in FY2022 for semiconductor-related policies.

As discussed in an earlier section, the development of semiconductors requires continuous investment. Therefore, the Japanese government is also considering a framework for continued support of the semiconductor industry. There are two frameworks: one based on the Economic Security Promotion Act, and the other a government bond-like framework tentatively named GX (Green Transformation) Economic Transition Bonds.

The Economic Security Promotion Act was passed by the Diet in May 2022, establishing a basic policy for ensuring the stable supply of designated critical components and materials. Semiconductors are one of such critical components. Private businesses can receive support by preparing a plan for the stable supply of such components and obtaining approval from the minister in charge.

As for the bond financing framework, the GX Implementation Council was launched in the Cabinet Secretariat in July 2022 to implement GX, which will transform the entire economic and social system by shifting its economic, social, and industrial structure from being fossil fuel-based to clean energy-based, as a green growth strategy to achieve carbon neutrality by 2050. To that end, discussions are underway to establish a bond-like framework for advancing GX and determine the budget amount in preparation for securing the necessary funding in the future. In this context, semiconductors are expected to be both a strategic component and one that promotes green growth.

(4) In addition, the company aims to achieve mass production in the late 2020s in cooperation with imec in Europe.

Imec is the world's most advanced research institute for semiconductors, employing 2700 researchers and 800 Ph.D. students. It is able to recruit students even though it does not award them degrees because it provides researchers with state-of-the-art prototype lines and evaluation equipment. It attracts researchers and students from the nearby Leuven Catholic University and around the world.

Technologies prototyped and evaluated at imec are highly valued and adopted around the world. Therefore, as many as 550 companies provide large amounts of joint research funding and send their top researchers to imec. About 80% of its total income of 420 million euros comes from foreign companies. This is invested in human resources and equipment.

Individuals who represent the interests of a consortium are not allowed to serve in management. A small number of executives visit the members in turns to flexibly and promptly respond to their requests, gather intelligence on market needs, and address them through a talented group of global personnel. In other words, the institute is equipped with a rich ecosystem to attract the brains of the world and innovate through co-existence and co-evolution (see Sect. 6.5, Column "The Secret of imec's Strength").

(5) What is required to create a data-driven society, Society 5.0, is advanced computing.

According to the "Fifth Science and Technology Basic Plan" of Japan's Cabinet Office, Society 1.0 is the hunting society, and Society 2.0 is the agricultural society. The primary industries of agriculture, forestry, and fisheries are labor-intensive, and diligence and persistence are the requirements for success.

Society 3.0 is the industrial society, while Society 4.0 is the information society. Manufacturing, the secondary industry, is capital-intensive, and "bigger is better" is the rule for success. Japan has become highly industrialized, but mass consumption is increasing the environmental burden, limiting growth, and widening the wealth gap. These are believed to be reasons for the population decline in developed countries.

The coming Society 5.0 is a human-centric society. The tertiary industry, the service industry, is knowledge-intensive, so sharing knowledge is important. It is a society where knowledge creates value, individuals can make the most of their individuality, and inclusiveness is advocated. Being inclusive means that all individuals have access to knowledge and equal opportunities.

(6) On the other hand, the power efficiency of the general-purpose processors used for that computation has improved by only one order of magnitude over the same decade.

Scaling reduces the capacitive component of the circuitry, thereby reducing power. Logic chips go through a generational change every two years, which can reduce power by about 30%. In other words, power can be reduced to $0.7^{10/2} \approx 0.17$ in 10 years. By adding design innovations to scaling, power efficiency has been improved by an order of magnitude.

The power consumed by the chip is released as heat, so if cooling cannot keep up, a portion of the circuit must be temporarily shut off. The part of the circuit that is shut off is called dark silicon. The percentage of dark silicon increases with scaling, reaching as high as 80% at 5 nm. In other words, even if more transistors can be integrated, their functions and performance may only be partially realized. Therefore, improvement in power efficiency can lead to improvement in performance.

Logic chips have gone through a generational shift every two years: 40, 28, 20, 16, 10, 7, 5, and 3 nm. But Japan is stagnant at 40 nm. TSMC's plant under construction in Kumamoto aims to manufacture chips at 16–28 nm. Compared to 16 nm, 2 nm is five generations ahead, and with power efficiency that is one order-of-magnitude higher. If a 2 nm chip can use the same power as a 16 nm chip, its performance can be 10 times higher. Alternatively, if used at the same performance, it will consume only one-tenth of the power. This is why Rapidus is targeting the most advanced 2 nm manufacturing process. In addition, it can be a strategy to catch up by skipping over the lost FinFET generations (16 to 3 nm) and taking on the challenge of GAA directly starting at 2 nm.

(7) If 3D integration can reduce the distance of data movement to a different order of magnitude, energy consumption in moving data will be greatly reduced.

When an object falls due to gravity, its potential energy is converted to kinetic energy, while friction and collision consume energy in the form of sound and heat. Similarly, when electrons move under an electric field, energy is consumed in the form of heat due to resistance along the path.

Scaling reduces the amount of charge required for calculations within a chip, but it does not reduce the amount of charge required to move data between chips. When chips encapsulated in separate packages are mounted on top of each other in the same package, the data transfer distance can be reduced by a factor of 10,000. As a result, the energy required to move data can be reduced by orders of magnitude.

Practical application will begin with stacking and assembling chips in the backend process, but eventually it will be possible to directly bond wafers and chips in the frontend process. The frontend and backend processes will gradually become one.

(8) However, it all changed with the emergence of flowering plants.

With the emergence of flowers, co-existence and co-evolution began between plants and insects. As a result, the forest ecosystem became richer, animal populations increased, and primates flourished (from "NHK Special: The Theory of Super-Evolution," The Message from Plants).

(9) Hundreds of billions of semiconductor switches in the audience's smartphones were turned on and off hundreds of millions of times.

Two types of switches that operate in opposite ways are connected between the power supply and ground to create a gate. When a low voltage ("zero") is input to the gate, the switch on the power supply side is turned on and that on the ground side is turned off, resulting in a high voltage ("one") output from the gate. Similarly, inputting "one" to the gate causes "zero" to be output from the gate.

When the input of a second gate is connected to the output of the first gate, and the output of the second gate is looped back to the input of the first gate, an input of "zero" to the first gate will store a "one" in its output. This creates a memory circuit. By properly designing the pathway, or circuit, through which electrons flow, we can either store or compute information.

In this way, by digitally processing digital information output by an image sensor, we can take pictures and then store them.

There are hundreds of billions of transistors integrated into a smartphone. If 1% of those transistors are used, then hundreds of billions of semiconductor switches will be working in 100 smartphones at the event. Also, if the switches toggle once every few clock cycles, where the clock ticks a billion times per second, the switches will be turning on and off hundreds of millions of times a second.

(10) Taro Hakase was playing 'Jounetsu Tairiku' in front of the audience.

"Jounetsu Tairiku" is the theme song of an old Japanese TV documentary series of the same name which profiled individuals who were pursuing their passions and making a significant impact in their fields.

2. Regaining Lost Ground: Game Changer

(11) However, it is difficult to make up for the time Japan lost in the last 30 years by conventional tactics alone. Therefore, it is also necessary to foresee the next battleground and make advanced investment accordingly ahead of the competition. This is referred to in the Japanese martial art of Kendo as 'sen-sen no sen o utsu,' or 'anticipate your opponent's next move and strike preemptively.

If you look at it from a philosophical point of view, the reason for why Japan's semiconductor industry fell behind and how to regain the lost ground can only be considered matters of national fortune. If the Japan–U.S. relationship changes from adversarial to cooperative, and the currency exchange from a strong to a weak yen, the climate will change from a headwind to a tailwind. We are now at a turning point in Japan's national fortune. Otto von Bismarck, who united Germany by force in the nineteenth century, asserted in a speech the importance of iron to a nation, which led to the expression "iron makes a nation." I wonder if in the twenty-first century semiconductors would become so important that people would say "semiconductors make a nation."

(12) In spite of that, in 15 years, Moore's law increased the level of integration by three orders of magnitude, and eventually, design could no longer keep up. Thus, the era of specialized chips came to an end.

Moore's Law is not a natural law. It is a description of a cadence of the industry.

A semiconductor is a giant collection of technologies. Meanwhile, the industry consists of a deep and wide supply chain from upstream to downstream. Therefore, the pace of technological development must be synchronized among many players. Instead of a three-legged race, it is more like a 20,000+1-legged race. Therefore, in order to synchronize their efforts, the semiconductor industry must first create a technology roadmap.

Even if manufacturer A outpaces competitor B in achieving the next process generational shift, its customers would not have anticipated and planned for it. In addition, since relying on a single manufacturer is risky, they often source from multiple suppliers. For these reasons, a particular cadence for process scaling has emerged in the industry.

The number of manufacturers capable of providing state-of-the-art technology has been decreasing recently. If such consolidation continues, the cadence will be disrupted. This is a sign of the end of Moore's Law.

(13) Optimism is an essential ingredient for innovation. How else can the individual welcome change over security, adventure over staying in safe place?

Regarding innovation, we may also find inspiration in another quote often attributed to Winston Churchill: "The pessimist sees difficulty in every opportunity. The optimist sees opportunity in every difficulty."

(14) By adopting this architecture, the evolution of the computer has followed a scenario where general-purpose hardware—processors and memory—are produced in large volumes to drive adoption, and software is used to tailor the hardware for various applications.

Computers have adopted the von Neumann architecture, which consists of a processor and memory. Therefore, the semiconductor market for processors and memory is huge. Mass production of standardized chips drives the pursuit of economic efficiency. This results in fierce competition in capital investment, which in turn has led to the current consolidation in the industry—Intel and Samsung Electronics account for 7 trillion yen of the processor and memory markets respectively.

(15) Japanese companies won the competition in device innovation but lost the competition in capital investment.

In 1988, Japan's share of the world semiconductor market was over 50%. Semiconductors were mainly used in consumer electronics such as televisions and video equipment. Japanese companies excelled at using analog technology to enhance the convenience of the physical space.
    Then the age of the PC and smartphone came, which created the virtual space through digital technology. It was an area where the U.S. excelled while Japan struggled.
    The future will require a sophisticated fusion of physical and virtual space. Japan is expected to be a leader in sensors and control motors which collect information from and drive the physical space respectively.
    In analog technology, quality improvement is pursued through the meticulous tuning and alignment of technology components. In digital technology, on the other hand, the focus is on scaling up by integrating modules. Japan is strong in the former, while the U.S. is strong in the latter. The difference in strength between the nations may be related to the difference in national character resulting from the contrast between Japan's homogeneous society and America's diverse society.

(16) In this manner, an era of general-purpose chips is started by device innovations and ended after fierce competition in capital investment. Meanwhile, an era of specialized chips is started by design methodology innovations and ended by Moore's Law.

For semiconductor manufacturers, profit is generated by mass production using the same set of photomasks. On the other hand, for semiconductor users, they create competitive products by securing custom-made chips tailored to their needs. As fierce market competition and technological innovation arise under these conflicting demands, the market swings between the eras of general-purpose and specialized chips like a pendulum.

As soon as the first mover in a new market starts to mass produce and make a profit, the number of entrants to the market increases rapidly, resulting in excessive competition that drives down prices and squeezes profits. Eventually, only those who have the resources to endure can survive the market consolidation.

On the other hand, those who lose the competition can turn to custom design technology to satisfy customer requirements. Therefore, those who can innovate in technology are the ones to open the door to the age of customization.

EDA (electronic design automation) was the technology innovation that launched the ASIC age in the 1980s. It countered the growing design complexity by increasingly raising the level of design abstraction from transistors to gates to logic.

However, even the best algorithm can realize performance improvement in problem-solving on the order of $n\log(n)$ at most. By contrast, Moore's Law increased the level of integration by a factor of 100 in 10 years, with which design innovations were unable to keep up. As a result, the age of product diversification came to an end and the market swung back to the era of general-purpose products.

The age of general-purpose chips continued from 2000 to 2020. But then the energy crisis of the data-driven society once again opened the door to the age of specialized chips. However, this time around, with Moore's law slowing down, the age of specialized chips may last a lot longer than before.

(17) Under such constraints, only those who can improve energy efficiency tenfold can achieve a tenfold increase in computing performance, or a tenfold increase in smartphone battery life.

If the capacitance of a circuit can be reduced by scaling, the energy consumption of the circuit can be reduced.

When the CMOS circuit turns on the switch on the power supply side, its capacitance is charged up to output a "one." And when the CMOS circuit turns on the switch on the ground side, the capacitance is discharged to produce a "zero."

When a voltage of V is applied to a capacitor with a capacitance of C, a charge of $Q = CV$ is stored. When that charge is moved from the power supply to ground, an energy of $E = QV$ is dissipated. In other words, the energy consumption of the CMOS circuit is $E = CV^2$.

The battery capacity of a smartphone is approximately 3000 mAh (milliampere-hours). The output voltage of a lithium battery is about 3.7 V. So 3 A $\times$ 3.7 V $\times$ 3600 s $= 40,000$ J of energy is stored.

Let's assume that the chip consumes 10 W of power for one second when taking a picture. In this case, 10 J of energy is consumed to take one photo, which means that the battery has the capacity for taking 4000 photos.

(18) The development of specialized chips is knowledge-intensive, not capital-intensive. Development of automatic layout and logic synthesis was previously driven by the University of California at Berkeley. Similarly, this time around, university research will play an integral role in creating the fundamental knowledge required for specialized chips development, including knowledge for automatic generation of functionalities and systems.

The problem of placing circuit blocks within a given area and routing between them according to specifications is an extremely complex combinatorial optimization problem.

If the number of blocks is 10, the number of possible combinations is the factorial of 10, i.e., 3.6 million plus, so the computer can exhaustively examine all the cases.

However, if the number of blocks is 100, the number of permutations exceeds 157 digits. In that case, it is necessary to give up on an exhaustive search and instead resort to methods that find solutions relatively close to the correct answer. Heuristic algorithms such as min-cut (minimum cut) and approximate computation methods such as simulated annealing have been used.

If the number of blocks is further increased to 1000, the number of permutations exceeds 2500 digits. That is a far bigger search space than the game of Go, which is said to have 360 digits of permutations. However, Google, which has released AI technology that outperforms even top players in Go, recently announced machine learning technology that can outperform experienced chip designers in place and route.

(19) The only way to solve society's energy problem is to increase semiconductor's energy efficiency. Compared to general-purpose chips, specialized chips can achieve a power efficiency that is two orders-of-magnitude better.

General-purpose chips are developed for general use and available on the open market. On the other hand, specialized chips are developed for specific applications and are not commercially available. Using processor as an example, Intel's CPUs and NVIDIA's GPUs are general-purpose chips, while Apple's M1 and Google's TPUs are specialized chips.

Because general-purpose chips are designed to be used by anyone for any purpose, there are always redundant circuits from the standpoint of any specific application. In addition, they must offer backward compatibility for product longevity, which means they are constrained by legacy requirements. In contrast, specialized chips can be optimally designed because the users and their purposes are well-defined. As a result, power efficiency can be increased by orders of magnitude.

(20) Soil is replaced by high-purity silicon.

Various materials besides high-purity silicon are used in semiconductors. In addition, the variety of materials used has been increasing rapidly in recent years. For example, wiring requires materials with higher electrical conductivity as scaling progresses. In the past, aluminum was used. But since 2000, copper and tungsten are also employed. In the future, we may need to switch to cobalt. Cobalt is a rare metal which is also used in lithium-ion batteries. The fact that the Democratic Republic of Congo in Africa accounts for more than half of the world's production is a source of concern for the supply chain.

3. Structural Transformation: More Moore

(21) There are three ways to reduce power—lowering voltage V, capacitance C, and the frequency of switching fa.

In electronic devices, information is carried by electrons. In the case of a CMOS circuit, the charge used for information processing is $Q = CV$ (C is the capacitance of the circuit and V the supply voltage). The energy consumed when this charge moves across a voltage V is $E = QV = CV^2$. Since power P is energy consumed per second, it is computed by multiplying energy with the frequency at which the circuit switches $P = faCV^2$ (f is the clock frequency and a is the switching probability).

(22) The manufacturing cost of a chip is computed by dividing the cost to manufacture a wafer by the number of good chips on the wafer.

Chip manufacturing cost is mainly determined by chip area, yield, and process complexity. The smaller the chip area and the higher the yield, the more good chips can be obtained from a single wafer. Also, the fewer the number of process steps, the lower the cost of producing a single wafer.

In addition to the cost of manufacturing the chip, there are additional costs for chip development, packaging, testing, and reliability testing, plus sales and marketing expenses. These costs, plus profit, determine the price.

(23) Device scaling is achieved through improvement to both lithography and process technology. At the same time, wafer size is increased while manufacturing technology is improved to increase yield, resulting in an increase in the number of good chips per wafer.

When rectangular chips are diced from a round wafer, the perimeter of the wafer is wasted. So why are wafers round?

The reason is that purity and homogeneity can be improved by spinning the wafer, whether it is during the manufacturing of high-purity single-crystal silicon ingots, resist application to wafers, layer deposition, or even cleaning of the wafers.

(24) If we reduce the supply voltage V [V] by the same factor of $1/\alpha$ as is used in shrinking the device dimension $\times$ [m] (a 20% shrink corresponds to $\alpha = 1.25$), the electric field inside the transistor [V/m] remains unchanged.

The unit of dimension is meter [m] and the unit of voltage is volt [V]. The unit for the strength of an electric field, or electric field intensity, is volt per meter [V/m]. Knowing the units of measurement is helpful in reminding us of the physical meaning.

A transistor that uses the electric field effect is called a field effect transistor (FET).

(25) Under this scaling scenario, both the current I [A] flowing through the transistor and its capacitance C [F] are also reduced by a factor of $1/\alpha$. This is because current is proportional to device dimension. On the other hand, capacitance is determined by area divided by thickness. Since area is reduced by $1/\alpha^2$ while thickness by $1/\alpha$, capacitance is reduced by $1/\alpha$.

The current I is the rate at which electric charge Q (unit is coulomb [C]) flows, so it is expressed in [C/s].

It is computed by multiplying two parameters—the density of charge along the length of the channel [C/m] induced by the electric field of the gate voltage, and the speed [m/s] with which charge is driven through the channel by the electric field between the drain and the source.

The charge density along the channel [C/m] is obtained by multiplying the gate capacitance C with the gate-channel voltage V, as expressed by $Q = CV$.

For a gate length of L, the gate capacitance is computed from $C = \varepsilon(LW)/d$. The capacitance along the channel is therefore determined by the channel width W [m]/gate dielectric thickness d [m]. In other words, the capacitance along the channel remains unchanged after scaling.

Therefore, the charge density along the channel is proportional to the voltage V, which is reduced to $1/\alpha$ by scaling.

On the other hand, the speed with which charge is driven through the channel is determined by the electric field between drain and source, i.e., the voltage between drain and source [V]/channel length [m], so the value remains unchanged after scaling.

In summary, the current I is proportional to V^2/x, so it is reduced to $1/\alpha$ by scaling.

The capacitance C is obtained from area/distance, so it is also reduced to $1/\alpha$ by scaling.

(26) When each of voltage, current, and capacitance scales down by the same factor of $1/\alpha$, circuit delay also scales down by $1/\alpha$. This is because delay is computed from the product of capacitance and voltage divided by current.

The load capacitance C of a rail-to-rail CMOS circuit with a supply voltage V is charged and discharged with an amount of charge given by $Q = CV$. The current I is the rate of charge flow, so $I = Q/t$ (t is time). Solving these two simultaneous equations yields $t = CV/I = RC$. In other words, the delay time of a CMOS circuit can be estimated by the RC time constant, which is the product of resistance R and capacitance C.

   Therefore, if the voltage V, current I, and capacitance C are all proportionally reduced by $1/\alpha$ respectively, the delay time of the CMOS circuit is also reduced by a factor of $1/\alpha$.

(27) Under this scenario, current I increases by a factor of α. Combined with a C reduced by a factor of $1/\alpha$, circuit delay decreases by a factor of $1/\alpha^2$. As a result, there is an additional increase in circuit speed. However, power density increases rapidly as a function of α^3, resulting in a proportional increase in heat generation.

From (25) to (26), if the dimension of the device is x [m], the current I [A] is proportional to V^2/x, the capacitance C [F] is proportional to x, and the circuit delay time [s] is proportional to CV/I. Therefore, if the dimension x of the device is reduced to $1/\alpha$ while the voltage V is kept constant, the current I increases by a factor of α with $V^2/x = 1^2/(1/\alpha)$, the capacitance C is reduced to $1/\alpha$ with $x = 1/\alpha$, and the delay time of the circuit is reduced to $1/\alpha^2$ with $CV/I = (1/\alpha)(1/\alpha)$, resulting in faster operation. However, the power density jumps by α^3 with $VI/x^2 = \alpha/(1/\alpha)^2$.

(28) As a result, impurities are implanted into the surface of the semiconductor substrate at the two ends of the gate, forming the source and drain.

Transistors and wiring have a three-dimensional structure. Using dozens of photomasks depicting the planar structure of each individual layer, the three-dimensional structure is created sequentially from the bottom to the top by repeatedly transferring the mask patterns layer by layer onto the chip surface. However, due to errors in the alignment of the masks, there will be slight offsets between the fabricated layers. The channel between the source and drain of the transistor must be as short as possible and directly under the gate. Therefore, a special manufacturing process is applied here to create the source, drain, and gate with highly precise locations. Specifically, the gate which is located above the source and drain is fabricated first, and then impurities are implanted from above. The gate, acting as a mask, prevents impurities from being implanted under the gate, while creating a source and drain that are perfectly aligned at the two ends of the gate. In this way, the pattern already formed in a previous step is used as a mask for the next step, allowing the fabrication to proceed without aligning the masks, in a process known as self-alignment.

(29) The next step taken was to change the transistor structure.

By changing the gate oxide layer to a material with a higher dielectric constant, it was possible to suppress the leakage current by increasing the oxide thickness while maintaining its capacitance. This is because leakage current is exponentially inversely proportional to the oxide thickness. This innovative approach to change the material of the gate oxide, which is at the heart of the transistor, was able to effectively suppress gate leakage current.

However, as scaling continued, junction leakage between the source and drain and the silicon substrate became significant, which finally required the transistor structure to be fundamentally changed.

(30) Meanwhile, Jack Kilby invented the integrated circuit (IC) in 1958. The challenge of wiring was overcome by using photolithography to simultaneously integrate many devices and wires onto a single chip. Eventually, silicon was found to be the best material for fabricating ICs.

Amber-colored LEDs with relatively long wavelengths are used for illumination in clean rooms to prevent photoresist for photolithography from becoming exposed. It is the same reason why a dark room is used in developing photographic films. As scaling continues, shorter wavelength light sources are used in exposure equipment. With the adoption of 13.5 nm EUV (extreme ultra-violet) light, white light illumination can now be used in clean rooms.

(31) However, as we entered the twenty-first century, with the successful creation of deep autoencoders and the achievement of sufficiently large computer performance required for training, deep learning became capable of delivering overwhelmingly larger processing performance than conventional processing, resulting in its rapid adoption.

A GPU (graphics processing unit) is a chip specialized for image processing. Various types of image processing (spatial filtering) can be applied by modifying the value of each pixel with the values of the neighboring pixels in different ways. For example, if the nine pixel values in a 3×3 region are each multiplied by a factor of 1/9 and added together, the pixels within the region are averaged, resulting in a blurred image. Conversely, if only the center pixel value is multiplied by 9 and the other pixel values are multiplied by $-$ 1 and added together (i.e., subtracted), the center pixel is enhanced, and the image is sharpened. Since all of these calculations can be expressed as matrix operations, GPUs are equipped with circuits that can efficiently perform matrix calculations.

In a neural network, an output signal is fired when the sum of the input signals to an axon multiplied by the synapse weights exceeds a threshold value. Therefore, neural network processing also involves repeated matrix operations. That is why GPUs are widely used for neural network computations.

(32) The Leading-edge Semiconductor Technology Center (LSTC) was established as a Collaborative Innovation Partnership (CIP).

CIP (collaborative innovation partnership) is a corporation established with the approval of the Minister of Economy, Trade and Industry. It enables multiple entities such as corporations, universities, and independent administrative institutions to collaborate on research and experimentation to overcome problems that no one entity can solve alone, and to put newly developed technologies to practical use.

CIP is an organization (a non-profit mutual benefit corporation) in which members conduct joint research on technologies needed for their industrial activities for their own, individual benefit. All members contribute researchers, research funds, equipment, and so on, to the joint research, jointly manage the results, and derive benefits from them together.

(33) They developed the reduction projection exposure system (also called a stepper) that dominated the global market and contributed to a jump in the domestic production ratio of semiconductor manufacturing equipment from 20 to 70%.

Japan is strong in materials and manufacturing equipment. In materials, Japan has a global share of over 65% of the 6 trillion yen market, and in manufacturing equipment, Japan has a 35% share of the 7 trillion yen market.

In the 1980s when it had a strong global presence in the semiconductor market, Japan leveraged its position as an integrated electronics manufacturer to develop strong technological capabilities in materials and manufacturing equipment. When its semiconductor industry eventually declined, it was able to maintain its competitiveness in the two areas by expanding its business globally. In addition, as is with the automotive industry, Japan has developed a deep and broad domestic ecosystem which it has maintained to this day.

Japan's strength is found in developing the implicit knowledge and know-how that is needed to find the optimal solution through experience and intuition to complex problems with many parameters: it is found where continuous improvement and refinement in the field is required. The development of new material products has a low probability of success, as described by the Japanese expression "sen-mitsu" that literally means "three in a thousand." In addition, the suppliers are expected to offer customization to tailor to the needs of each customer. In short, passion for manufacturing excellence, strict quality control, patience in development, and the readiness to thoroughly address customer needs are all characteristics of the Japanese people attributable to their competitiveness.

Exploration of materials informatics (MI), or the use of the computer to explore materials, has begun. It is not clear yet if MI will be a threat or a blessing to Japan. Some suggest that MI will make the strong companies even stronger.

(34) To that end, we need to mobilize everyone to join the academic pursuit of diverse disciplines from design to device, manufacturing, equipment, and materials, in order to achieve total optimization instead of piecemeal, partial optimizations.

As scaling becomes more and more difficult and transistor structure changes radically to FinFET and GAA, design-technology co-optimization (DTCO), and even further, system-technology co-optimization (STCO) are becoming necessary.

4. A Fertile Ground for Innovation: More than Moore

(35) Without being able to rely solely on on-chip integration for performance increase, we have been evolving from 2 to 3D chip integration, which calls for a breakthrough solution to the connection problem.

Logic chips, DRAM, and NAND flash have very different requirements for the semiconductor device. Logic chips require transistors that operate at high speed; DRAM requires capacitors with low leakage; and NAND requires an ultra-thin layer that can trap electrons.

For example, if you make logic chips and DRAM on the same wafer, you must build both high-speed transistors and high-performance capacitors. If the logic chips and DRAM occupy about the same area, the use of a process for high-performance capacitors in the logic region is a waste; likewise, the use of a process for high-speed transistors in the DRAM region is a waste.

As a result, manufacturing logic and DRAM as separate chips and then mounting them close together reduces costs compared to integrating them on the same chip.

(36) With TCI, communication is achieved by using the digital signal to be transmitted to control the direction of current flow in coils formed on the interconnect layers of the transmitting chip to alter the direction of the associated magnetic field, which in turn changes the polarity of the signal induced in matching coils on the receiving chip. The induced signal is then detected and used to regenerate the original digital signal in the receiver.

Coils have long been used in oscillators of analog circuits. In order to minimize parasitic capacitance and resistance as much as possible, the layout area has become relatively large, and rarely are more than 10 of them placed on a single chip.

On the other hand, in the case of TCI, since it is a digital circuit, it is acceptable to have some parasitic capacitance or resistance. Multilayer wiring allows for smaller circuit layout areas. Wires can pass through coils, and circuits can be placed under coils. More than 1000 coils can be placed on a single chip.

(37) If the brain is connected to the internet, and its ability to innovate grows as the number of people connected increases, will ideas reproduce so fast as to completely overwhelm the Earth, like the way Matt Ridley described in his book The Rational Optimist?

The Internet has made instantaneous spatial travel possible: online meetings allow you to travel the world. However, the problem of time difference remains. Online meetings with participants from around the world are often held late at night or early in the morning.

So, could we also cross the time barrier if we could connect our brains to the Internet? For meetings in the middle of the night, we could go to sleep as usual. When it is time to meet, brain waves would be induced to allow the participants to attend while in the REM stage of sleep. During REM sleep, the body rests with skeletal muscles relaxed, but the brain is active and awake. In fact, the brain is more active than during the day and the mind is clearer. Discussions would probably be more productive.

The problem is, come morning, you would not remember what you said at the meeting. You may be surprised at what you said when you look at the meeting minutes. If that happens, you could literally blame it on you being half asleep.

5. Democratization: More People

(38) For instance, if we compare the computational complexity of sorting algorithms for an input size of n, quick sort which employs 'divide and conquer' reduces complexity to $O(n\log_2 n)$ compared to bubble sort which has a complexity of $O(n^2)$.

The divide-and-conquer approach can also reduce the time required in search from $O(n)$ for linear search to $O(\log_2 n)$ for binary search.

(39) Johannsen's adviser was Mead.

You can watch a speech by Johannsen in celebration of Mead's 80th birthday that is filled with love for his mentor at the following site: https://www.youtube.com/watch?v=9kz1ZWO1Dr8.

As an aside, I have an anecdote to tell about the need for care in layout color assignments. In the US, where Mead's textbook was used, red was assigned to the polysilicon gate. But at Toshiba where I worked, red was for aluminum interconnects, so it caused a lot of confusion.

(40) I was a visiting scholar at UC Berkeley in 1988. My host was Séquin who developed RISC-1 together with David Patterson.

Professor Carlo Séquin's research career has been colorful: he graduated in experimental physics at the University of Basel, Switzerland, in 1965 and became a pioneer in CCD development at Bell Laboratories in the U.S. He then moved to UC Berkeley in 1977, where he developed the world's first RISC processor with Professor Patterson. He has been working on CAD and computer graphics research since 1984, and has since expanded into the adjacent fields of architecture and visual arts.

I once said to him, "Your way of life is just like simulated annealing."

In Newton's method for numerical analysis, for example, the solution space search only proceeds in the direction of improving evaluation value. Therefore, depending on the initial value, you may end up at a local instead of the global optimal point. On the other hand, simulated annealing is different in that sometimes the search proceeds in a direction of deteriorating evaluation value, with a probability that depends on the temperature. Just like annealing, or tempering, the search bounces around the search space a lot when the temperature is high, but as the temperature is gradually lowered it settles down.

I thought that approach could also be applied to life. But instead of settling down, Prof. Séquin continues to bounce around. "You never seem to cool down at all," I said. "In fact, you seem to be heating up instead." Then, he gave me a mischievous look.

Simulated annealing is suitable for exploring static space, but the space I'm interested in is dynamic and changing.

6. Epilogue: Super-Evolution

(41) While 'bottom' is the exploration of nano (10^{-9}) dimensions, 'top' is the exploration of massive integration in giga ($10^9 = 1$ billion) and even tera ($10^{12} = 1$ trillion) quantities.

Integrated circuits are described with numbers that range from the minuscule to the gigantic. First, prefixes for indicating small numbers include micro (10^{-6}), nano (10^{-9}), pico (10^{-12}), femto (10^{-15}), and atto (10^{-18}). Micro to nano are used for device feature size (in units of meters); pico to femto are used for capacitance (in units of farad); pico to femto are also often used for signal propagation time (in units of seconds). Micro to milli (10^{-3}) are used for current (in units of amperes), while kilo ($10^3 = 1000$) to mega ($10^6 = 1$ million) are often used for resistance (in units of ohms).

On the other hand, prefixes for indicating large numbers include giga ($10^9 = 1$ billion), tera ($10^{12} = 1$ trillion), peta ($10^{15} = 1$ quadrillion), and exa ($10^{18} = 1$ quintillion). Giga to tera are used for memory and storage capacities (in units of bytes); giga to tera are also used for the speed of data transfer between chips (in units of bits/second); peta to exa are used for supercomputer processing performance (in units of operations/second). The total amount of digital data in circulation (in units of bytes) exceeds exa.

Thus, integrated circuits deal with an astronomical span of space that measures a whopping 36 zeros.

Glossary of Useful Terms to Know

Architecture Basic design and design philosophy of a computer.

ASIC Abbreviation for Application-Specific Integrated Circuit, an integrated circuit for a specific application, i.e., a specialized chip.

Backend Processing Process of wiring and encapsulating a chip in a package.

CAD Abbreviation for Computer-Aided Design, a computer tool that assists design.

Chip A semiconductor integrated circuit about 1 cm^2 in size with transistors and wiring integrated on a silicon substrate.

CMOS Abbreviation for Complementary Metal–Oxide–Semiconductor, a circuit in which P-type and N-type transistors operate in a complementary manner.

Compile Conversion of source code written in a programming language into a machine language that can be directly executed by a computer.

CPU Abbreviation for Central Processing Unit, a chip that performs data processing.

Devices Components of electronic circuits such as transistors and wiring.

DRAM Abbreviation for Dynamic Random Access Memory, a memory chip that temporarily stores data.

EDA Abbreviation for Electronic Design Automation, which refers to the automation of design work for semiconductors and electronic devices, or its tools and software.

EUV Lithography Abbreviation for Extreme Ultraviolet Lithography, which is the most advanced exposure technology using extreme ultraviolet light with short wavelengths.

FinFET Abbreviation for Fin Field-Effect Transistor, a transistor with a three-dimensional structure that increases its gate strength compared to older transistors built on the surface of a chip, whose name is derived from its resemblance to the fin of a fish.

Flash A memory chip that stores data for a long period of time, available in NAND and NOR types.

Foundry A company that specializes in the manufacturing of chips developed by design companies.

© The Editor(s) (if applicable) and The Author(s), under exclusive license to Springer Nature Switzerland AG 2025
T. Kuroda, *The Super-Evolution of Semiconductors*, Synthesis Lectures on Engineering, Science, and Technology, https://doi.org/10.1007/978-3-031-60518-5

FPGA Abbreviation for Field-Programmable Gate Array, an integrated circuit whose circuits can be programmed post-production.

Frontend Processing Process of manufacturing devices on wafers.

GAA Abbreviation for Gate-All-Around, a new type of transistor in which the gate surrounds the channel to further increase gate strength compared to FinFET.

Gate Control terminal to turn on/off a transistor.

General-purpose Chip A chip that is sold commercially and used for general purposes, such as memory chips and Intel CPUs.

GPU Abbreviation for Graphics Processing Unit, a chip that excels at parallel processing and is suited for graphics and AI processing.

Imec Abbreviation for Interuniversity Microelectronics Centre, the Belgian research institute which leads the world in microfabrication technology.

Logic Chip A chip that processes data.

Moore's Law A rule of thumb stating that chip integration doubles every 1.5–2 years.

Memory Chip A chip that stores data.

Photomask An original plate used in photolithography to transfer elements and circuit patterns onto a silicon substrate, dozens of which are used to manufacture a chip.

Photolithography A technique for transferring a pattern from a photomask onto a chip which consists of resist coating, exposure, and development processes.

Process Chip manufacturing process.

Semiconductors Substances that have properties between conductors that conduct electricity and insulators that do not conduct electricity, which can be used to regulate the flow of electricity.

Scaling The process of continuous reduction of the size of manufactured devices on an integrated circuit.

SoC Abbreviation for System on a Chip, a chip designed to function as a system by integrating a processor core, microcontroller and specialized functions on a single chip.

Specialized Chip A chip that is not sold in the marketplace and is used for a specific purpose, such as Google's AI chips and Apple's CPUs.

Transistor A semiconductor device capable of amplifying or switching electrical signals.

TSMC Abbreviation for Taiwan Semiconductor Manufacturing Company, the world's largest foundry headquartered in Taiwan.

VLSI Abbreviation for Very Large-Scale Integration, a complex and large-scale chip integrating more than 100,000 transistors

Wafer A thin disk made from slicing a cylinder of single-crystal silicon, used as the base material for chip manufacturing.